北京理工大学"双一流"建设精品出版工程

Illustration of the Cutting-edge Modern Biotechnology
(English-Chinese Bilingual Version)

图解现代生物技术前沿
（英汉对照版）

主　编　◎　霍毅欣
副主编　◎　马晓焉

北京理工大学出版社
BEIJING INSTITUTE OF TECHNOLOGY PRESS

内容简介

本教材是一本专注于生物技术前沿领域的教学参考书,涵盖了生命科学与技术的前沿发展方向,并介绍了生物技术在不同领域的最新应用。教材内容以顶级期刊的综述文献、诺贝尔奖等主流奖项的成果为参考,采用卡通图示来阐述科学原理与技术,以英文原文为蓝本并加以了中文注释。全书分为七章,由多位教授领衔编写,涉及通用生物技术、工业生物技术、农业生物技术、环境生物技术、谱学检测技术、肿瘤治疗分析技术以及生物信息技术等多个方面。本书旨在开阔本科高年级本科生和研究生的科研视野,有助于提高科技英文阅读及生命科学研究水平。

版权专有　侵权必究

图书在版编目(CIP)数据

图解现代生物技术前沿:英汉对照 / 霍毅欣主编. -- 北京:北京理工大学出版社,2019.12
　　ISBN 978-7-5682-8031-0

Ⅰ. ①图… Ⅱ. ①霍… Ⅲ. ①生物工程-图解-英、汉 Ⅳ. ①Q81-64

中国版本图书馆 CIP 数据核字(2019)第 292723 号

责任编辑:李颖颖　　　　　**文案编辑**:辛丽莉
责任校对:周瑞红　　　　　**责任印制**:李志强

出版发行 /	北京理工大学出版社有限责任公司
社　　址 /	北京市丰台区四合庄路 6 号
邮　　编 /	100070
电　　话 /	(010)68914026(教材售后服务热线)
	(010)68944437(课件资源服务热线)
网　　址 /	http://www.bitpress.com.cn

版 印 次 /	2019 年 12 月第 1 版第 1 次印刷
印　　刷 /	三河市华骏印务包装有限公司
开　　本 /	787 mm×1092 mm　1/16
印　　张 /	18.75
彩　　插 /	8
字　　数 /	346 千字
定　　价 /	58.00 元

图书出现印装质量问题,请拨打售后服务热线,负责调换

前　言

生物技术有着悠久的发展和应用历史，从早期的酿造工艺到当前的基因编辑，生物技术的施用对象已由生物个体深入到分子水平，其应用范围也从传统的农业与食品工业逐步扩展到环保、医疗、制造业等领域，为人类应对粮食安全、气候变化、生命与健康、资源能源安全等重大挑战提供了崭新的解决方案。目前，生物技术已成为多种相关学科的基础必修课程。

然而，本书编者在生物技术前沿领域的教学中感受到了一些问题。第一，生物技术导论类课程的对口教材缺失，内容差异明显，难以横向评估教师的教学和学生的学习质量。第二，教材以文字性资料为主，不容易理解。生物技术并非基础理论科学，而是由技术方案、技术规程、应用案例等设计性、流程性内容组成的应用科学，以传统的文字方式叙述的教材不易于学生理解。第三，教学内容落后于学科发展前沿。生物技术发展迅猛，现已由对生命规律的初级应用发展为对生命的人工再造，大多数教学内容仍限于对传统，甚至是过时技术的介绍，对当前生物技术的最新进展和应用领域，以及未来的发展方向关注不足。

本书顺应了国内生物技术相关专业的迫切需求，全书聚焦生命科学与技术的前沿发展方向，以及在其他领域的最新应用；以顶级期刊的最新综述文献、诺贝尔奖等主流奖项的成果为参考；以卡通图示阐述科学原理与技术；以英文原文为蓝本并加以中文注释，减小了教材的难度。本书将为"生命科学基础""生物工程与技术导论""生物学与生物技术前沿""现代生命科学与技术述评"等课程提供参考材料，有助于开阔本科高年级学生以及研究生的科研视野，帮助其建立科研思路，提高其科技英文阅读及生命科学的研究水平。

本书共分7章，由霍毅欣教授领衔编写。第1章为通用生物技术，解释了常用生物技术的理论原理和技术流程，由霍毅欣教授和马晓焉副研究员编写；第2章为工业生物技术，包括合成途径设计与工程菌株构建内容，附加典型应用案例，由霍毅欣教授编写；第3章为农业生物技术，介绍了生物技术在生物肥料、生物农药、作物育种等方面的前沿应用，由马晓焉副研究员编写；第4章为环境生物技术，介绍了包括土壤污染、水环境污染、大气污染、固体废弃物污染在内的生物处理技术，由杨宇副教授编写；第5章介绍了光谱、色谱、质谱等谱学检测技术及相关仪器的工作原理和最新进展，由徐伟教授编写；第6章为肿瘤治疗分析技术，介绍了肿瘤的特征及其微环境，肿瘤的研究方法，以及小分子药物和免疫治疗等肿瘤治疗技术，由董磊教授编写；第7章为生物信息技术，介绍了高通量测序技术、生物信息数据分析技术的最新进展以及应用，由金花教授编写。

　　本书的编写和图表绘制离不开团队诸多成员的辛勤劳动，同时也得到了夏琴、翟雁冰、唐丹、张佩、毋彤、马炼杰、于盛竹、茹家康、宋威等师生的鼎力协助，在此一并表示衷心的感谢！

　　受限于作者团队的专业背景及研究领域，本书难免存在疏漏以及不当之处，在此表示歉意，也恳请各位同行及同学们包涵并指正。

编　者

Preface

Biotechnology has a long history of development and application. From the early brewing to the current gene editing, the biotechnology object has deepened from the individual level to the molecular level, and the application fields has also expanded from the traditional agriculture and food industry to environmental protection, medicine, manufacturing industry and other fields, providing new solutions for human beings to deal with major challenges such as food security, climate changes, life and health, and resource and energy security. At present, biotechnology has become a basic required course for many related disciplines.

However, the authors of this book have experienced some problems in the teaching of cutting-edge biotechnology. First, there is a lack of books that are specifically designed for introductory biotechnology courses, and the significant differences in contents make it difficult to horizontally evaluate the teaching quality of teachers and the learning quality of students. Second, the teaching materials are mainly textual materials. Biotechnology is not a basic theoretical science, but an applied science composed of technical solutions, operating procedures and application cases of design and procedural contents. Traditional books that are narrated in a textual way are not easy for students to understand. Third, the contents lag behind the frontier of discipline development. The rapid developing biotechnology has evolved from the primary application of the laws of life to the artificial reconstruction of life. Most teaching materials are still limited to the introduction of traditional and even outdated technologies, with little attention paid to the latest advances and application areas of current biotechnology, as well as future developing directions.

This book is in line with the urgent needs of relevant majors in China. It focuses on the cutting-edge development direction of life science and technology, as well as the latest applications of biotechnology in other fields. The contents are based on the latest review literature of top journals, the achievements of mainstream awards such as the Nobel Prize, and the scientific principles and technologies are explained with cartoon illustrations, and in both English and Chinese. This book will provide reference materials for courses such as "Basics of Life Science" "Introduction to Bioengineering and Technology" "Frontiers of Biology and Biotechnology" "Review of Modern Life Science and Technology", which will help broaden the scientific research horizons of senior undergraduates and graduate students, help them establish scientific research ideas, and improve their technical English reading ability and professional level in life science.

This book is issued under Prof. Huo Yixin general authorship and is divided into seven chapters. The 1st chapter written by Prof. Huo Yixin and associate researcher Ma Xiaoyan introduces general biotechnology, which explains the the oretical principles and processes of commonly used techniques; the 2nd chapter written by Prof. Huo Yixin introduces industrial biotechnology, including the construction of the design of synthetic pathways, and engineering strains, with additional typical application cases; the 3rd chapter written by associate researcher Ma Xiaoyan is agricultural biotechnology, which introduces the frontier applications of biotechnology in biofertilizers, biopesticides, crop breeding, etc.; the 4th chapter written by associate professor Yang Yu introduces environmental biotechnology, including biological treatment technologies of soil pollution, water environment pollution, air pollution, and solid waste pollution; the 5th chapter written by Prof. Xu Wei introduces the working principle and latest progress of spectral detection technologies such as spectroscopy, chromatography, mass spectrometry and related instruments; the 6th chapter written by Prof. Dong Lei introduces tumor treatment analysis technology, including the characteristics of tumors and their microenvironment, research methods of tumors, as well as tumor treatment technologies such as small molecule drugs and immunotherapy; the 7th chapter written by Prof. Jin Hua introduces the leading edge of bioinformatics technology and its applications, including the latest progress of high-throughput sequencing technology and bioinformatics data analysis technology.

The compilation and illustration of this book is inseparable from the hard work of

Preface

many team members as well as the strong assistance of teachers and students such as Xia Qin, Zhai Yanbing, Tang Dan, Zhang Pei, Wu Tong, Ma Lianjie, Yu Shengzhu, Ru Jiakang, Song Wei, etc. I would like to express my heartfelt gratitude together!

Limited by the professional backgrounds and research fields of the author team, this book will inevitably have negligence or inappropriateness. We apologize for this, and hope our colleagues and classmates would forgive and correct us.

Authors

目 录

第1章 通用生物技术 ………………………………………………………………… 1
Chapter 1　General Biotechnology ………………………………………………… 1
  Ⅰ　基因编辑技术 ……………………………………………………………………… 4
  Ⅰ　Gene Editing Technology ………………………………………………………… 4
    1.1　基因编辑技术的演变 ………………………………………………………… 4
    1.1　The Evolution of Gene Editing Technology ……………………………… 4
    1.2　CRISPR/Cas9 技术原理 ……………………………………………………… 6
    1.2　Principle of CRISPR/Cas9 Technology …………………………………… 6
    1.3　CRISPR/Cas9 的应用 ………………………………………………………… 8
    1.3　Applications of CRISPR/Cas9 ……………………………………………… 8
  Ⅱ　基因线路 …………………………………………………………………………… 10
  Ⅱ　Genetic Circuit …………………………………………………………………… 10
    2.1　基因线路的逻辑门 …………………………………………………………… 10
    2.1　Logic Gate of Genetic Circuit ……………………………………………… 10
    2.2　代谢工程中的可编程基因线路 ……………………………………………… 12
    2.2　Programmable Genetic Circuits in Metabolic Engineering ……………… 12
    2.3　基因路线在不同领域中的应用 ……………………………………………… 13
    2.3　Application of Genetic Circuits in Different Fields ……………………… 13
  Ⅲ　定向进化 …………………………………………………………………………… 15
  Ⅲ　Directed Evolution ……………………………………………………………… 15
    3.1　定向进化流程 ………………………………………………………………… 15
    3.1　Workflow of Directed Evolution …………………………………………… 15
    3.2　加速定向进化的策略 ………………………………………………………… 17
    3.2　Strategies for Accelerating Directed Evolution …………………………… 17
    3.3　定向进化的前景 ……………………………………………………………… 18
    3.3　Prospects for Directed Evolution …………………………………………… 18
  Ⅳ　适应性实验室进化 ………………………………………………………………… 19
  Ⅳ　Adaptive Laboratory Evolution ………………………………………………… 19
    4.1　适应性实验室进化方案 ……………………………………………………… 19
    4.1　Scheme of Adaptive Laboratory Evolution ……………………………… 19

4.2　适应性实验室进化实施策略 ·· 20
　　4.2　Strategies for Effective Adaptive Laboratory Evolution ············· 20
Ⅴ　蛋白质设计 ··· 22
Ⅴ　**Protein Design** ··· 22
　　5.1　蛋白质工程技术 ·· 22
　　5.1　Protein Engineering Techniques ·· 22
　　5.2　天然与人工蛋白的合成 ··· 23
　　5.2　Synthesis of Natural and Artificial Proteins ··· 23
Ⅵ　分子模拟 ··· 23
Ⅵ　**Molecular Simulation** ··· 23
　　6.1　分子动力学模拟 ·· 23
　　6.1　Molecular Dynamics Simulation ·· 23
　　6.2　分子模拟的过程及功能 ··· 25
　　6.2　Process and Function of Molecular Simulation ··· 25
　　6.3　分子模拟的应用实例 ·· 26
　　6.3　Application Examples of Molecular Simulation ··· 26
Ⅶ　结构生物学 ··· 28
Ⅶ　**Structural Biology** ··· 28
　　7.1　结构生物学的研究对象及工具 ·· 28
　　7.1　Research Objects and Tools of Structural Biology ····································· 28
　　7.2　揭示生物分子系统结构信息的方法 ··· 29
　　7.2　Methods for Revealing Structural Information of Biomolecular Systems ········ 29
　　7.3　蛋白质数据库 ··· 30
　　7.3　The Protein Data Bank ·· 30
Ⅷ　生物传感器 ··· 32
Ⅷ　**Biosensors** ··· 32
　　8.1　生物传感器的定义与应用 ··· 32
　　8.1　Definition and Application of Biosensors ··· 32
　　8.2　生物传感器在高通量筛选中的潜在应用 ··· 34
　　8.2　Potential Applications of Biosensors in High-throughput Screening ············· 34
Ⅸ　高通量筛选技术 ··· 36
Ⅸ　**High-throughput Screening Technology** ·································· 36
　　9.1　高通量筛选方案 ·· 36
　　9.1　High-throughput Screening Strategies ·· 36
　　9.2　基于FACS的高通量筛选 ··· 38
　　9.2　FACS-based High-throughput Screening ·· 38
Ⅹ　酶固定化 ··· 40
Ⅹ　**Enzyme Immobilization** ··· 40
Ⅺ　无细胞催化 ··· 42
Ⅺ　**Cell-free Catalysis** ··· 42

11.1	无细胞催化技术及其关键流程	42
11.1	Cell-free Catalytic Technology and Key Processes	42
11.2	无细胞系统来源、优化及应用	45
11.2	CFS Source, Optimization and Application	45
XII	人工混菌体系	47
XII	Synthetic Microbial Consortia	47

第2章 工业生物技术 49
Chapter 2 Industrial Biotechnology 49

I 原料的利用 52
I Utilization of Raw Materials 52

1.1	工业生产原料的种类及发展趋势	52
1.1	Types of Raw Materials for Industrial Production and Their Development Trends	52
1.2	工业原料利用概况	54
1.2	Overview of Industrial Raw Materials Utilization	54
1.3	蛋白质的利用及代谢途径	63
1.3	Utilization and Metabolic Pathways of Proteins	63
1.4	简单糖的利用及代谢途径	67
1.4	Utilization and Metabolic Pathways of Simple Sugar	67
1.5	木糖的利用及代谢途径	69
1.5	Utilization and Metabolism Pathways of Xylose	69
1.6	木质纤维素的利用及代谢途径	71
1.6	Utilization and Metabolic Pathways of Lignocellulose	71
1.7	CO_2 的利用及代谢途径	73
1.7	Utilization and Metabolic Pathways of CO_2	73
1.8	原料利用能力的强化	74
1.8	Enhancement of Raw Material Utilization	74

II 产物的合成 76
II Product Synthesis 76

2.1	生物燃料合成	76
2.1	Biofuel Synthesis	76
2.2	芳香族化合物合成	77
2.2	Aromatic Compound Synthesis	77
2.3	天然产物合成	78
2.3	Natural Product Synthesis	78
2.4	聚合物合成	82
2.4	Polymer Synthesis	82

III 底盘的改造 83
III Modification of Microbial Chassis 83

| 3.1 | 常见底盘 | 83 |
| 3.1 | Common Chassis | 83 |

  3.2 通过代谢工程进行底盘改造 ……………………………………………… 85
  3.2 Chassis Modification Through Metabolic Engineering ……………… 85
  3.3 微生物底盘的优化 ………………………………………………………… 87
  3.3 Optimization of Microbial Chassis ……………………………………… 87
  3.4 理想细菌底盘的特征 ……………………………………………………… 89
  3.4 Characteristics of Ideal Bacterial Chassis ……………………………… 89
  3.5 提高产量的改造策略 ……………………………………………………… 90
  3.5 Engineering Strategies to Improve Production ………………………… 90
  3.6 新型生物活性化合物的挖掘 ……………………………………………… 92
  3.6 Mining of New Bioactive Compounds …………………………………… 92

第3章 农业生物技术 ……………………………………………………………… 95
Chapter 3 Agricultural Biotechnology …………………………………………… 95

 Ⅰ 微生物技术 …………………………………………………………………………… 97
 Ⅰ Microbial Biotechnology …………………………………………………………… 97
  1.1 土壤健康 …………………………………………………………………… 97
  1.1 Soil Health ………………………………………………………………… 97
  1.2 作物生产 …………………………………………………………………… 101
  1.2 Crop Production …………………………………………………………… 101
  1.3 肠道微生物与动物健康 …………………………………………………… 108
  1.3 Intestinal Microbes and Animal Health ………………………………… 108
 Ⅱ 植物生物技术 ………………………………………………………………………… 119
 Ⅱ Plant Biotechnology ………………………………………………………………… 119
  2.1 增产 ………………………………………………………………………… 119
  2.1 Yield Increase ……………………………………………………………… 119
  2.2 抗虫作物 …………………………………………………………………… 125
  2.2 Insect Resistant Crops …………………………………………………… 125
  2.3 作物品质提升 ……………………………………………………………… 126
  2.3 Crop Quality Improvement ……………………………………………… 126

第4章 环境生物技术 …………………………………………………………………… 129
Chapter 4 Environmental Biotechnology ………………………………………… 129

 Ⅰ 环境生物技术的定义 ………………………………………………………………… 131
 Ⅰ Definition of Environmental Biotechnology ……………………………………… 131
 Ⅱ 环境生物技术的内涵及研究方法 …………………………………………………… 132
 Ⅱ The Connotation and Research Methods of Environmental Biotechnology …… 132
  2.1 环境生物技术的特点 ……………………………………………………… 132
  2.1 Characteristics of Environmental Biotechnology ……………………… 132
  2.2 基于系统生物学的应用方法 ……………………………………………… 134
  2.2 Application Methods Based on Systems Biology ……………………… 134
 Ⅲ 前沿进展及应用技术 ………………………………………………………………… 136
 Ⅲ Frontier Progress and Application Technology …………………………………… 136

3.1	污染物的生物传感及检测	136
3.1	Biosensing and Detection of Pollutants	136
3.2	海洋石油污染微生物降解及修复	149
3.2	Microbial Degradation and Repair of Marine Oil Pollution	149
3.3	可降解塑料的生物生产及合成塑料的生物降解	155
3.3	Bioproduction of Biodegradable Plastics and Biodegradation of Synthetic Plastics	155
3.4	厌氧氨氧化及废水处理的应用	161
3.4	Application of Anaerobic Ammonium Oxidation and Waste Water Treatment	161
3.5	合成生物学在环境生物技术中的应用	166
3.5	Application of Synthetic Biology in Environmental Biotechnology	166

第5章　光谱、色谱及质谱检测技术 173
Chapter 5　Detection Technology of Spectrum, Chromatography and Mass Spectrometry 173

Ⅰ　光谱技术 175
Ⅰ　Spectral Technique 175

1.1	光谱仪器的介绍	175
1.1	Introduction to Spectral Instruments	175
1.2	光谱仪器的分类	177
1.2	Classification of Spectral Instruments	177

Ⅱ　色谱技术 185
Ⅱ　Chromatographic Technique 185

2.1	色谱技术概述	185
2.1	Overview of Chromatographic Technique	185
2.2	气相色谱技术	187
2.2	Gas Chromatography	187
2.3	高效液相色谱技术	190
2.3	High Performance Liquid Chromatography	190

Ⅲ　质谱技术 194
Ⅲ　Mass Spectrometry Technique 194

3.1	质谱分析	194
3.1	Mass Spectrometry	194
3.2	常见质谱仪的分类及介绍	199
3.2	Category and Introduction of Common Mass Spectrometers	199

第6章　肿瘤治疗分析技术 209
Chapter 6　Treatment and Analysis Technology of Tumors 209

Ⅰ　肿瘤的特征及其微环境 211
Ⅰ　Characteristics and Microenvironment of Tumor 211

1.1	肿瘤的特征	211
1.1	The Nature of Tumor	211
1.2	肿瘤的发展	212

1.2 Tumor Development ········· 212
  1.3 肿瘤微环境 ········· 213
  1.3 Tumor Microenvironment ········· 213
 Ⅱ 肿瘤研究方法 ········· 217
 Ⅱ Tumor Research Methods ········· 217
  2.1 建立肿瘤动物模型 ········· 217
  2.1 Establishment of Tumor Animal Model ········· 217
  2.2 肿瘤生物信息学 ········· 220
  2.2 Tumor Bioinformatics ········· 220
 Ⅲ 肿瘤的治疗 ········· 225
 Ⅲ Treatment of Tumors ········· 225
  3.1 小分子靶向治疗 ········· 225
  3.1 Small Molecule Targeted Therapy ········· 225
  3.2 免疫治疗技术 ········· 228
  3.2 Immunotherapy Techniques ········· 228

第7章 生物信息技术 ········· 233
Chapter 7 Bioinformatics Technology ········· 233

 Ⅰ 高通量测序技术 ········· 235
 Ⅰ High-throughput Sequencing Method ········· 235
  1.1 高通量测序方法与原理 ········· 236
  1.1 High-throughput Sequencing Method and Its Principle ········· 236
  1.2 文库构建方法和应用 ········· 246
  1.2 Library Construction Method and Application ········· 246
 Ⅱ 生物信息学分析方法和理论 ········· 260
 Ⅱ Bioinformatics Analysis Methods and Theories ········· 260
  2.1 基因组从头测序的数据处理 ········· 260
  2.1 Data Processing of De Novo Genome Sequencing ········· 260
  2.2 全基因组重测序的数据处理 ········· 265
  2.2 Data Processing of Whole-genome Resequencing ········· 265
  2.3 宏基因组测序的数据处理 ········· 272
  2.3 Data Processing of Metagenomic Sequencing ········· 272
  2.4 RNA 测序数据处理 ········· 275
  2.4 Data Processing of RNA Sequencing ········· 275

参考文献 ········· 281

第1章 通用生物技术

Chapter 1 General Biotechnology

本章提要

通用生物技术
- 基因编辑技术
  - 基因编辑技术的演变
  - CRISPR/Cas9技术原理
  - CRISPR/Cas9的应用
- 基因线路
  - 基因线路的逻辑门
  - 代谢工程中的可编程基因线路
  - 基因线路在不同领域中的应用
- 定向进化
  - 定向进化流程
  - 加速定向进化的策略
  - 定向进化的前景
- 适应性实验室进化
  - 适应性实验室进化方案
  - 适应性实验室进化实施策略
- 蛋白质设计
  - 蛋白质工程技术
  - 天然与人工蛋白的合成
- 分子模拟
  - 分子动力学模拟
  - 分子模拟的过程及功能
  - 分子模拟的应用实例
- 结构生物学
  - 结构生物学的研究对象及工具
  - 揭示生物分子系统结构信息的方法
  - 蛋白质数据库
- 生物传感器
  - 生物传感器的定义与应用
  - 生物传感器在高通量筛选中的潜在应用
- 高通量筛选技术
  - 高通量筛选方案
  - 基于FACS的高通量筛选
- 酶固定化 —— 酶固定化方法及其应用
- 无细胞催化
  - 无细胞催化技术及其关键流程
  - 无细胞系统来源、优化及应用
- 人工混菌体系 —— 混菌体系的应用

Summary of Chapter 1

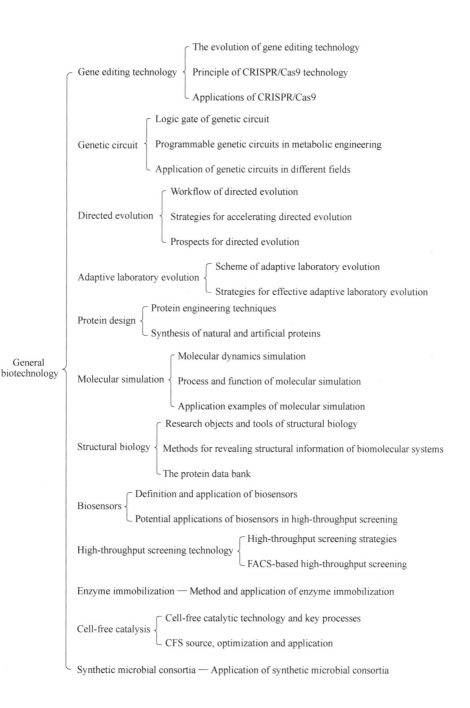

1 基因编辑技术
1 Gene Editing Technology

1.1 基因编辑技术的演变
1.1 The Evolution of Gene Editing Technology

图 1-1-1 所示为基因编辑技术发展时间线。

Fig. 1-1-1 shows the timeline of gene editing technology.

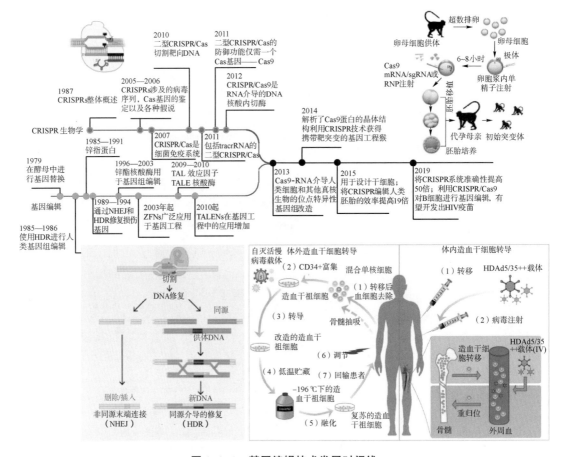

图 1-1-1 基因编辑技术发展时间线

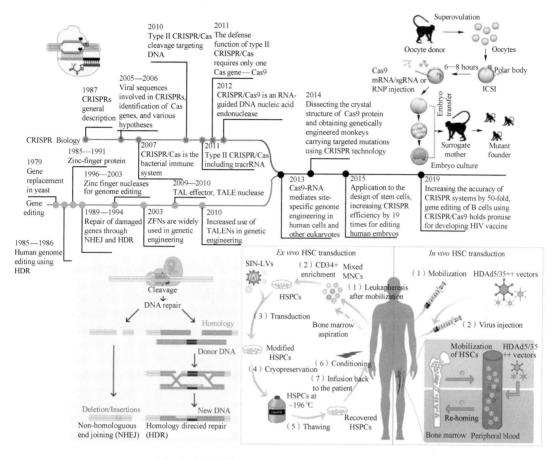

Fig. 1-1-1　Timeline of cene editing technology

注释：
HDR，同源重组修复；NHEJ，非同源末端连接；ZFNs，锌指核酸酶；TALENs，转录激活子样效应因子核酸酶；CRISPR，规律性重复短回文序列簇。

Note：
HDR, homologous recombination repair; NHEJ, non-homologous end joining; ZFNs, zinc finger nucleases; TALENs, transcription activator-like effector nucleases; CRISPR, clustered regularly interspaced short palindromic repeats.

1.2 CRISPR/Cas9 技术原理
1.2 Principle of CRISPR/Cas9 Technology

图 1-1-2 所示为 CRISPR/Cas9 技术原理。

Fig. 1-1-2 shows the principle of CRISPR/Cas9 technology.

CRISPR/Cas9 系统是一种基因组改造工具。图 1-1-2 (a) 所示为将平末端双链 DNA 断裂引入基因组位点的不同策略。这些位点成为内源性细胞 DNA 修复机制的靶标，催化非同源末端连接或同源介导修复。如图 1-1-2 (b) 所示，当 Cas9 的组氨酸-天冬酰胺-组氨酸结构域或 RuvC 结构域活性位点上包含失活突变点时，它可以作为一种切口酶 (nCas9) 发挥作用。当 nCas9 与两个能够识别目标位点附近 DNA 的 sgRNA 一起使用时，就会产生交错双链断裂。如图 1-1-2 (c) 所示，当 Cas9 的两个活性位点都含有失活突变时，它便成为 RNA 引导的 DNA 结合蛋白。特别是当融合了抑制结构域或激活结构域时，这种无催化活性或失活的 Cas9 (dCas9) 就可以介导转录下调或激活。此外，dCas9 可以融合如绿色荧光蛋白 (GFP) 的荧光结构域，用于染色体位点的活细胞成像。其他 dCas9 融合，如包括染色质或 DNA 修饰结构域的融合，可能使基因组 DNA 发生靶向表观遗传学改变。

CRISPR/Cas9 system is a genome engineering tool. Fig. 1-1-2 (a) shows different strategies for introducing blunt double-stranded DNA breaks into genomic loci. These loci become substrates for endogenous cellular DNA repair machinery that catalyze non-homologous end joining or homology directed repair. As shown in Fig. 1-1-2 (b), Cas9 can function as a nickase (nCas9) when engineered to contain an inactivating mutation in either the HNH domain or RuvC domain active sites. When nCas9 is used with two sgRNAs that recognize offset target sites in DNA, a staggered double-strand break is created. As shown in Fig. 1-1-2 (c), Cas9 functions as an RNA-guided DNA binding protein when engineered to contain inactivating mutations in both of its active sites. This catalytically inactive or dead Cas9 (dCas9) can mediate transcriptional down-regulation or activation, particularly when fused to repressor or activator domains. In addition, dCas9 can be fused to fluorescent domains, such as green fluorescent protein (GFP), for live-cell imaging of chromosomal loci. Other dCas9 fusions, such as those including chromatin or DNA modification domains, may enable targeted epigenetic changes in genomic DNA.

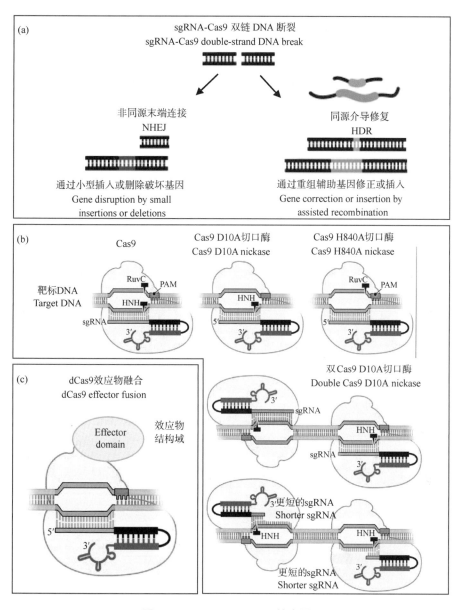

图 1-1-2　CRISPR/Cas9 技术原理

Fig. 1-1-2　Principle of CRISPR/Cas9 technology

(a) 将平末端双链 DNA 断裂引入基因组位点；(b) Cas9 作为一种切口酶发挥作用；

(c) Cas9 成为 RNA 引导的 DNA 结合蛋白

(a) Introducing blunt double-stranded DNA breaks into genomic loci; (b) Cas9 can function as a nickase;

(c) Cas9 functions as an RNA-guided DNA binding protein

1.3 CRISPR/Cas9 的应用
1.3 Applications of CRISPR/Cas9

图 1-1-3 所示为可使用 Cas9 进行改造的细胞类型和物种及 CRISPR/Cas9 的应用前景。

Fig. 1-1-3 shows examples of cell types and organism that have been engineered using Cas9 and the future applications of CRISPR/Cas9.

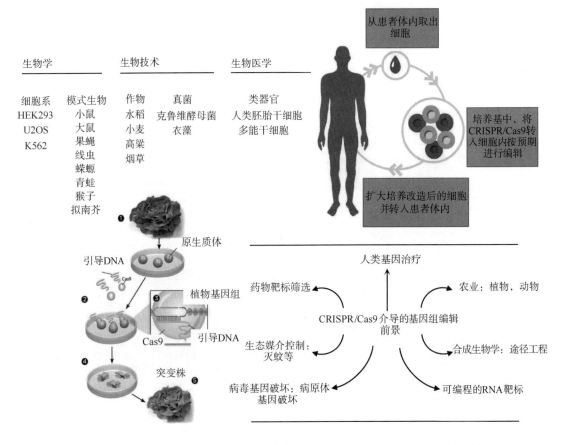

图 1-1-3　可使用 Cas9 进行改造的细胞类型和物种及 CRISPR/Cas9 的应用前景

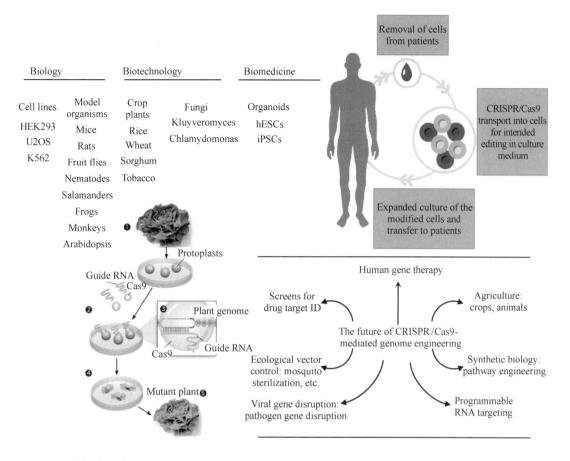

Fig. 1-1-3 Examples of cell types and organisms that have been engineered using Cas9 and the future applications of CRISPR/Cas9

2 基因线路
2 Genetic Circuit

2.1 基因线路的逻辑门
2.1 Logic Gate of Genetic Circuit

用原核生物转录调节模块可构建逻辑门模型。图1-2-1所示为各种控制启动子活性的元件：RNA聚合酶、转录因子、转录因子结合位点和相互作用等。每一个逻辑门模型均由转录因子结合位点（操纵基因）和RNA聚合酶结合位点（启动子）组装而成。一个操纵基因和一个启动子的重叠会引起转录抑制，而置于启动子上游的操纵基因会引起转录激活。图1-2-1（a）所示的"与"门可以通过一个依赖于诱导物B的转录因子A来激活启动子实现。图1-2-1（b）所示的"或"门可以是一种由两个独立的转录因子完全激活的启动子。图1-2-1（c）所示的"非"门相当于转录抑制。

Models of logic gates can be built with prokaryotic transcription regulatory modules. Fig. 1-2-1 shows various actors that control promoter activity: RNA polymerase (RNAP), transcription factors, binding DNA sites and interactions. Each of the gate models are assembled by combination of TFs binding sites (operators) and RNAP binding sites (promoters). The overlapping of one operator with a promoter causes repression while operators placed upstream from the promoter causes activation. The AND-gate, as shown in Fig. 1-2-1 (a) can be implemented as a TF that depends on an inducer B to activate the promoter. The OR-gate, as shown in Fig. 1-2-1 (b) could be a promoter amenable to full activation by two independent TFs. The NOT-gate, as shown in Fig. 1-2-1 (c) is equivalent to transcriptional repression.

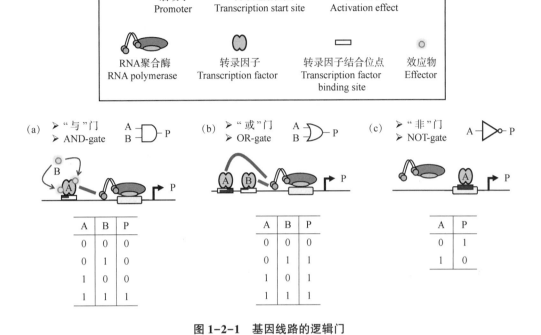

图 1-2-1　基因线路的逻辑门

Fig. 1-2-1　Logic gate of genetic circuit

2.2 代谢工程中的可编程基因线路
2.2 Programmable Genetic Circuits in Metabolic Engineering

细胞可感知其环境条件和代谢状态。这些输入通过复杂而强大的基因线路，如图1-2-2所示，为有关基因调控的"决策"提供信息。多基因靶向策略将允许同时和正交调控不同的基因靶标。在此情况下，低温和低氧允许固氮酶的表达，这种酶可以固氮但会被氧气不可逆地灭活。同时，糖原合成将被下调，以确保有足够的葡萄糖可为固氮过程提供能量。固定的氮可激活蓝藻素合成酶，它将固定的氮储存为蓝藻素，从而增加固氮酶的通量。真值表描述了由一组给定的环境和代谢输入信号产生的预期输出。

Cells can sense their environmental conditions and metabolic state. These inputs inform "decisions" regarding gene regulation via a complex and robust genetic circuit, as shown in Fig. 1-2-2. Multi-gene targeting strategies will allow for the simultaneous and orthogonal regulation of distinct gene targets. In this case, low temperature and low oxygen allow the expression of nitrogenase, an enzyme that can fix nitrogen but is irreversibly inactivated by oxygen. At the same time, glycogen synthesis will be down-regulated to ensure that there is sufficient glucose available to provide energy to the process of nitrogen fixation. Fixed nitrogen activates the expression of cyanophycin synthase, which stores fixed nitrogen as cyanophycin, and increase flux through nitrogenase. The truth table describes the expected outputs that will result from a given set of environmental and metabolic input signals.

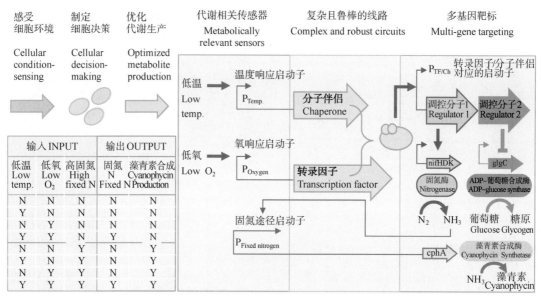

图 1-2-2 代谢工程应用中的可编程基因线路

Fig. 1-2-2 Programmable genetic circuit in a metabolic engineering application

2.3 基因路线在不同领域中的应用
2.3 Application of Genetic Circuits in Different Fields

图 1-2-3（a）所示为一个控制柴油替代品生产的回路，该回路显示通过对糖的感应调节有毒中间体的累积，从而诱发羟甲基戊二酸单酰辅酶 A 还原酶（HMGR）的周期性合成。如图 1-2-3（b）所示，可以通过检测单核苷酸多态性（SNP）并将该信息与组织特异性传感器整合，从而构建基于 CRISPRi 技术的基因治疗线路。图 1-2-3（b）展示了一个可以检测两个结肠癌易感性相关 SNP 的线路，通过整合结肠细胞特异性启动子，可以控制失调基因 NO3 和 DDX28。如图 1-2-3（c）所示，控制细菌定植人体菌群并执行治疗反应。一个设想是利用共生细菌稳定的 pH 值来疗胃食管酸反流。可通过基因线路改造天然的胃部共生细菌实现 pH 值的稳定，该线路的控制器输出质子泵抑制剂，从而将 pH 控制在设定值。如图 1-2-3（d）所示，基因线路也可用于构建能够感知环境刺激并做出反应的 "智能植物"。这里，我们设想了一个可在叶绿体中运行的线路，通过感应干旱、温度和植物成熟度以实现多性状控制。这可以降低进入食品供应的重组蛋白的合成量而不削弱其功能性。

Fig. 1-2-3 (a) shows a circuit that controls the production of diesel substitutes regulates the accumulation of toxic intermediates through sugar sensing to induce the periodic synthesis of hydroxymethylglutaryl-CoA reductase (HMGR). As shown in Fig. 1-2-3 (b), gene therapy circuits could be built based on CRISPRi technology by detecting SNPs and integrating this information with tissue specific sensors. Fig. 1-2-3 (b) shows a circuit that could detect two SNPs associated with colon cancer susceptibility and this is integrated with a promoter that is specific to colon cells to control the expression of misregulated genes, NO3 and DDX28. As shown in Fig. 1-2-3 (c), bacteria could be programmed to colonize human microbiota and implement a therapeutic response. An example is envisioned where a commensal bacterium is used to stabilize pH to treat gastoesophageal acid reflux. A bacterium that is naturally commensal with the stomach could be programmed to maintain the pH using a circuit that enables set point control via a controller whose output is proton pump inhibitors. As shown in Fig. 1-2-3 (d), genetic circuits could also be used to build "smart plants" that are able to sense environmental stimuli and implement a response. Here, we envision a circuit that would operate in the chloroplast integrate sensors for drought, temperature, and plant maturity to control multiple traits. This could reduce the amount of recombinant protein that is produced and enters the food supply without reducing the effectiveness of the trait.

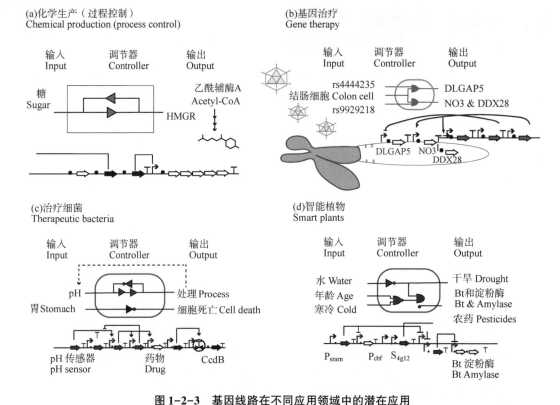

图 1-2-3 基因线路在不同应用领域中的潜在应用

Fig. 1-2-3 Potential use of genetic circuits in different application areas

(a) 一个控制柴油替代品生产的回路;(b) 构建基于 CRISPRi 技术的基因治疗线路;
(c) 控制细菌定植人体菌群并行治疗反应;
(d) 基因线路也可用于构建能够感知环境刺激并做出反应的"智能植物"

(a) A circuit that controls the production of diesel substitutes regulates the accumulation of toxic intermediates;

(b) Gene therapy circuits could be built based on CRISPRi technology;

(c) Bacteria could be programmed to colonize human microbiota and implement a therapeutic response;

(d) Genetic circuits could be used to build "smart plants"

Ⅲ 定向进化
Ⅲ Directed Evolution

3.1 定向进化流程
3.1 Workflow of Directed Evolution

定向进化流程如下：图 1-3-1（a）所示为在待改造的酶的基因中引入随机突变；在图 1-3-1（b）中，将基因引入细菌，细菌以基因文库为模板产生带有随机突变的酶；在图 1-3-1（c）中，对改造后的酶进行测试，筛选出能催化目标反应的最高效突变体；在图 1-3-1（d）中，向筛选出的酶中引入新的随机突变，重复上述步骤。

The workflow of directed evolution is as below: In Fig. 1-3-1 (a), random mutations are introduced in the gene for the enzyme that will be changed, In Fig. 1-3-1 (b), the genes are introduced into bacteria, and bacteria use the gene library as templates to produce enzymes with random mutations; In Fig. 1-3-1 (c), the changed enzymes are tested. Those that are the most efficient at catalyzing the desired reaction are selected; In Fig. 1-3-1 (d), new random mutations are introduced in the gene for the selected enzyme. The cycle begins again.

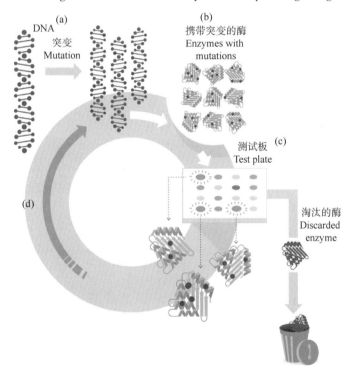

图 1-3-1 定向进化的一般流程
Fig. 1-3-1 General workflow for directed evolution
(a) 待改造的酶的基因中引入随机突变；(b) 细菌以基因文库为模板产生带有随机突变的酶；
(c) 筛选出能催化目标反应的最高效突变体；(d) 向筛选出的酶中引入新的随机突变
(a) Random mutations are introduced in the gene for the enzyme;
(b) Bacteria use the gene library as templates to produce enzymes with random mutations;
(c) Those that are the most efficient at catalyzing the desired reaction are selected;
(d) New random mutations are introduced in the gene for the selected enzyme

尽管突变空间是多维的，将定向进化视为三维适应度地貌中的一系列步骤有助于概念理解。产生的库反映了适应度地貌表面的大体情况，而筛查或选择指明了通往最高适应度的遗传途径。定向进化可以达到绝对的最高活性水平，但当库的多样性不足以跨越"适应度谷"而通往相邻的适应度峰值时，定向进化也会停留在局部的最高适应度水平。图 1-3-2 所示为定向进化过程的三维空间示意。

Although the mutational space is multidimensional, it is conceptually helpful to visualize directed evolution as a series of steps within a three-dimensional fitness landscape. Library generation samples the proximal surface of the landscape, and screening or selection identifies the genetic means to "climb" toward fitness peaks. Directed evolution can arrive at absolute maximum activity levels but can also become trapped at local fitness maxima in which library diversification is insufficient to cross "fitness valleys" and access neighboring fitness peaks. Fig. 1-3-2 shows three-dimensional spatial representation of the directed envolutionary process.

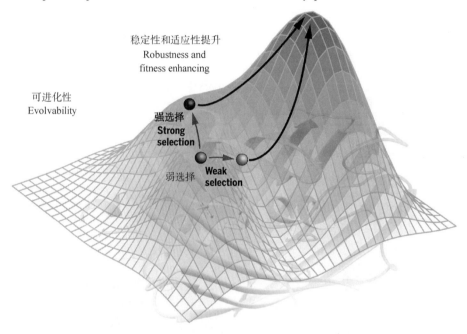

图 1-3-2　定向进化过程的三维空间示意

Fig. 1-3-2　Three-dimensional spatial representation of the directed evolutionary process

3.2 加速定向进化的策略
3.2 Strategies for Accelerating Directed Evolution

三种合成生物学工具可以拓宽筛查或选择指标范围。如图 1-3-3（a）所示，新型生物传感器可以检测目标代谢物并触发筛查或选择标记的表达；如图 1-3-3（b）所示，基因线路可以提高输入信号（目标表型）的复杂度，并将其关联至目标输出（可筛查或可选择的表型）；如图 1-3-3（c）所示，噬菌体辅助连续进化（PACE）可将目标性状与噬菌体适应度的提升相关联，从而实现多轮连续进化。

Three distinct synthetic biology tools can broaden the spectrum of screenable or selectable traits. As shown in Fig. 1-3-3 (a), novel biosensors can allow detection of metabolites of interest and trigger expression of screenable or selectable markers; As shown in Fig. 1-3-3 (b), genetic circuits can increase the complexity of inputs (targeted phenotype) that can be connected to a desired output (screenable or selectable phenotype); As shown in Fig. 1-3-3 (c), phage-assisted continuous evolution (PACE) can correlate the trait of interest with improvements in phage fitness, enabling multiple rounds of continuous evolution.

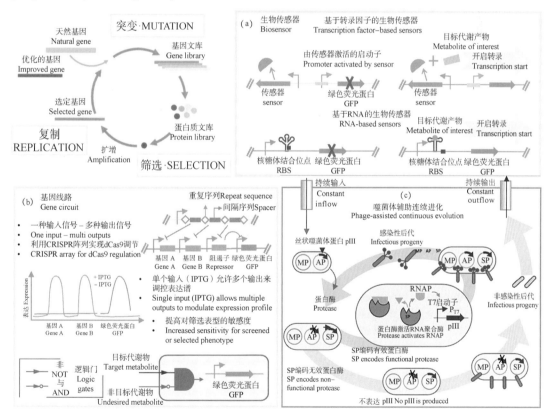

图 1-3-3 通过高效的筛查和选择策略加速定向进化
Fig. 1-3-3 Accelerating directed evolution through efficient screening and selection strategies
（a）新型生物传感器检测目标代谢物并触发筛查或选择标记的表达；
（b）基因线路提高输入信号的复杂度，并将其关联至目标输出；（c）噬菌体辅助连续进化实现多轮连续进化
（a）Novel biosensors allow detection of metabolites of interest and trigger expression of screenable or selectable markers；
（b）Genetic circuits increase the complexity of inputs that can be connected to a desired output；
（c）Phage-assisted continuous evolution can correlate the trait of interest, enabling multiple rounds of continuous evolution

3.3 定向进化的前景
3.3 Prospects for Directed Evolution

图 1-3-4 所示为定向进化的应用前景。由图 1-3-4（a）可见，随着蛋白质计算设计方法向更可靠的从头方法发展，合成生物学的所有领域将迎来新的可能。图 1-3-4（b）所示为依托定向进化方法产生的化合物。

Fig. 1-3-4 shows application prospects for directed evolution. From Fig. 1-3-4（a）we can see, as computational protein design methods advance towards more reliable denovo methods, new possibilities will be opened in all areas of synthetic biology. Fig. 1-3-4（b）shows compounds produced by directed evolutionary approaches.

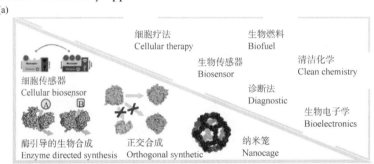

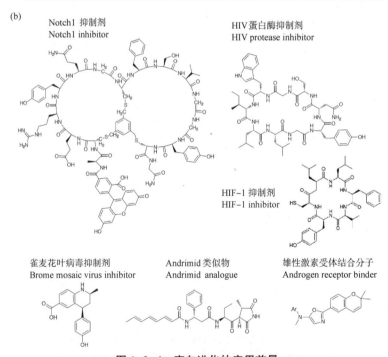

图 1-3-4　定向进化的应用前景

Fig. 1-3-4　Application prospects of directed evolution

（a）合成生物学的所有领域将迎来新的可能；（b）依托定向进化方法产生的化合物

（a）New possibilities will be opened in all areas of synthetic biology;

（b）Compounds produced by directed evolutionary approaches

Ⅳ 适应性实验室进化
Ⅳ Adaptive Laboratory Evolution

4.1 适应性实验室进化方案
4.1 Scheme of Adaptive Laboratory Evolution

首先，体内诱变方法打开了一扇新的大门，在高突变率和菌株增殖的条件下连续引入突变使基因型空间最大化并加速适应性实验室进化（ALE）。其次，最新的生长偶联策略使ALE能够增强更多有吸引力的性状（如非天然底物同化、生产促进）。此外，多重自动化培养平台可更精准地控制筛选，可独立复制并降低劳动力成本，为获得预期微生物铺平了道路。最后，基因组、转录组、蛋白质组和代谢组等多组学分析，以及多重基因工程的应用，有助于高效获取微生物代谢和基因调控的相关知识。图1-4-1所示为适应性实验室进化方案。

First, *in vivo* mutagenesis methods open a new gate to maximize the genotypic space and accelerate adaptative laboratory evolution (ALE) by continuously introducing mutations at a high mutation rate with strain proliferation. Then, the up-to-date growth-coupling strategies enable ALE to improve upon a wider range of appealing traits (e.g., non-native substrate assimilation, production improvement). Furthermore, multiplexed automated culture platforms pave the way to obtain desired microbes with more precise control of selection, more independent replicates, and less labor cost. Finally, the application of multi-omics analyses, including genomic, transcriptomic, proteomic, and metabolomic analyses, and multiplexed genetic engineering facilitate the efficient knowledge mining of microbial metabolism and gene regulations. Fig. 1-4-1 shows scheme of adaptive laboratory evolution.

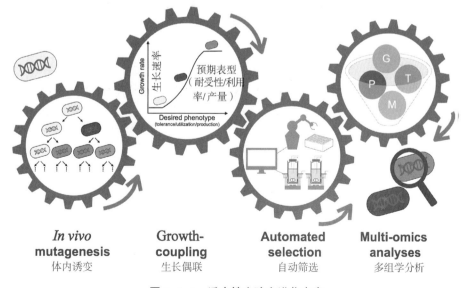

图 1-4-1 适应性实验室进化方案
Fig. 1-4-1 Scheme of adaptive laboratory evolution

4.2 适应性实验室进化实施策略
4.2 Strategies for Effective Adaptive Laboratory Evolution

图 1-4-2 所示为可用于适应性实验室进化的前沿生长偶联策略。图 1-4-2（a）为代谢生长偶联策略。通过敲除旁路途径，可将目标途径设计为生物量生产的必需途径。因此，可通过 ALE 增强靶标途径，从而有助于增强底物同化能力或产物生产等性状；图 1-4-2（b）所示为合成生物传感器引导的生长偶联策略。在操作范围内，生物传感器可以通过检测胞内浓度或胞外实际浓度来促进高产个体的生长；图 1-4-2（c）所示为量身定制的生长偶联策略。可以利用特定的化学品或物理生长条件来促进相关产物的生产。富含稀有密码子的抗生素抗性标记和噬菌体辅助的连续进化也具有应用潜力。

Fig. 1-4-2 shows advanced growth-coupling strategies for effective adaptive laboratory evolution. Fig. 1-4-2 (a) shows the metabolic growth-coupling strategy. By knockout of bypass pathways, the targeted pathway is designed to be the essential pathway tied with biomass generation. Therefore, the targeted pathway could be enhanced through ALE, contributing to traits like enhanced substrate assimilation or product production. Fig. 1-4-2 (b) shows synthetic biosensor-guided growth-coupling strategy. Within the operational range, biosensors can aid the growth of overproducers by sensing intracellular concentration or real extracellular titer. Fig. 1-4-2 (c) shows tailor-made growth-coupling strategy. Specific chemicals or physical growth conditions could be exploited to enhance the production of related products. Rare codon-rich antibiotic-resistant markers and phage-assisted continuous evolution also show their potential.

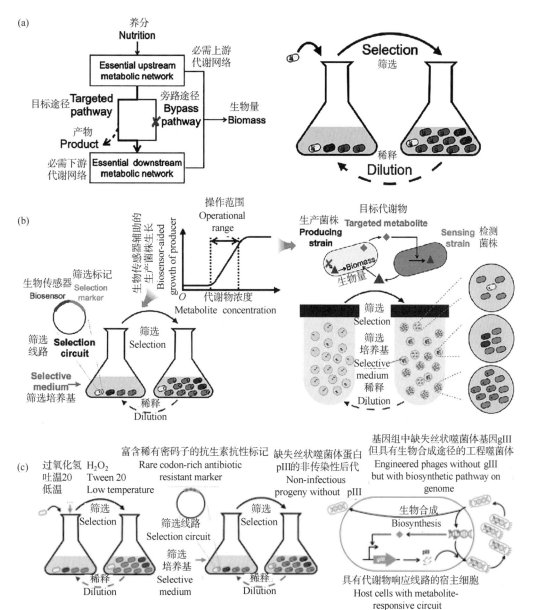

图 1-4-2 可用于适应性实验室进化的前沿生长偶联策略

Fig. 1-4-2 Advanced growth-coupling strategies for effective adaptive laboratory evolution

（a）代谢生长偶联策略；（b）合成生物传感器引导的生长偶联策略；（c）量身定制的生长偶联策略

(a) Metabolic growth-coupling strategy; (b) Synthetic biosensor-guided growth-coupling strategy;

(c) Tailor-made growth-coupling strategy

V 蛋白质设计
V Protein Design

5.1 蛋白质工程技术
5.1 Protein Engineering Techniques

图 1-5-1 所示为蛋白质工程技术。

Fig. 1-5-1 shows the protein engineering techniques.

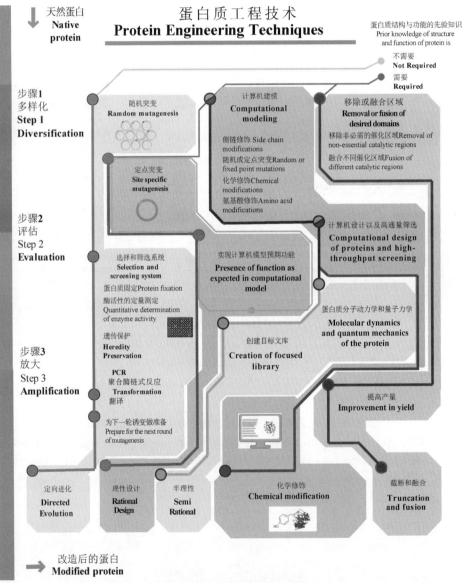

图 1-5-1 蛋白质工程技术

Fig. 1-5-1 Protein engineering techniques

5.2 天然与人工蛋白的合成
5.2 Synthesis of Natural and Artificial Proteins

图 1-5-2 所示为天然与人工蛋白的合成过程。

Fig. 1-5-2 shows synthesis of natural and artificial proteins.

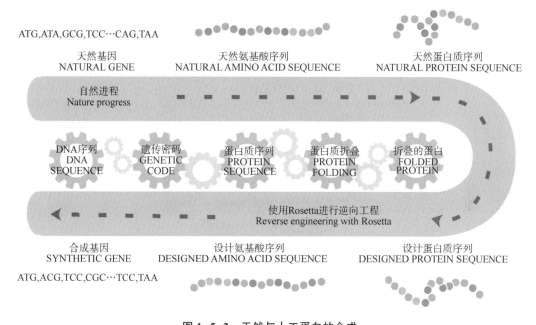

图 1-5-2 天然与人工蛋白的合成

Fig. 1-5-2 Synthesis of natural and artificial proteins

Ⅵ 分子模拟
Ⅵ Molecular Simulation

6.1 分子动力学模拟
6.1 Molecular Dynamics Simulation

分子模拟是一种利用计算机，以原子水平的分子模型来模拟分子结构与行为，进而模拟分子体系的物理与化学性质的方法。它是在实验基础上，通过基本原理构筑起一套模型和算法，从而计算出合理的分子结构与分子行为。分子模拟不仅可以模拟分子的静态结构，也可模拟分子体系的动态行为。分子模拟的主要方法有两种：分子蒙特卡罗法和分子动力学法。分子模拟的工作可分为两类：预测型和解释型。预测型工作是对材料进行性能预测、对过程进行优化，进而为实验提供可行的解决方案。解释型工作即通过现象模拟、理论建立和机理探索，为实验奠定理论基础。图 1-6-1 所示为分子模拟的内容。

Molecularsimulation, a method of simulating molecular structures and behaviors at the atomic level, uses a computer to simulate the various physical and chemical properties of a molecular system. Based on experiments and through the basic principles, it constructs a set of models and algorithms to calculate reasonable molecular structure and molecular behavior. Molecular simulation can not only simulate the static structure of a molecule, but also simulate the dynamic behavior of a molecular system. There are two main methods of molecular simulation: molecular Monte Carlo and molecular dynamics. The work of molecular simulation can be divided into two categories: predictive and interpretative. The predictive work is to predict the performance of the material, optimize the process, and provide a feasible solution for the experiment. Interpretative work lays a theoretical foundation for experiments by simulating phenomena, establishing theory, and exploring mechanisms. Fig. 1-6-1 shows contents of molecular simulation.

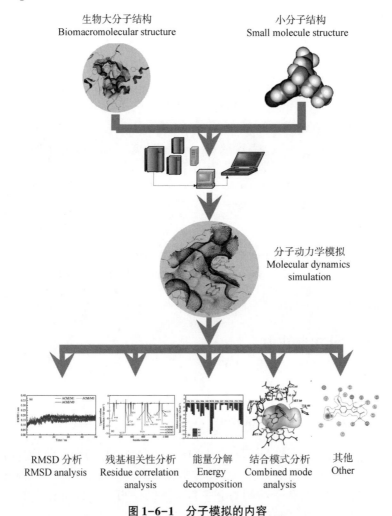

图 1-6-1 分子模拟的内容

Fig. 1-6-1 Contents of molecular simulation

6.2 分子模拟的过程及功能
6.2 Process and Function of Molecular Simulation

图 1-6-2 所示为分子模拟的过程及功能。

Fig. 1-6-2 shows the process and function of molecular simulation.

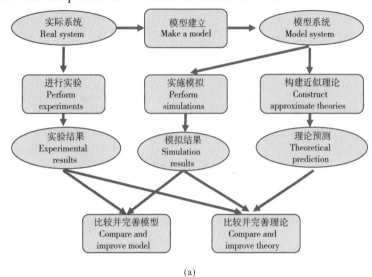

(a)

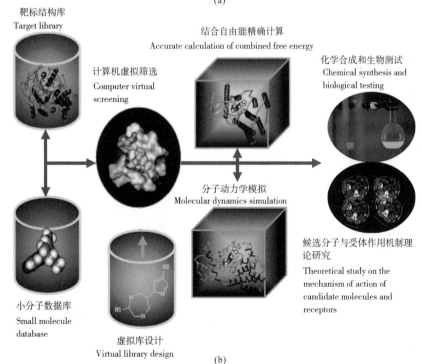

(b)

图 1-6-2 分子模拟的过程及功能

Fig. 1-6-2 Molecular simulation process and function

(a) 分子模拟的过程；(b) 分子模拟的功能

(a) Molecular simulation process; (b) Molecular simulation function

6.3 分子模拟的应用实例
6.3 Application Examples of Molecular Simulation

图 1-6-3 所示为分子动力学模拟的一些普遍应用实例。

Fig. 1-6-3 shows some of the common applications of MD simulations.

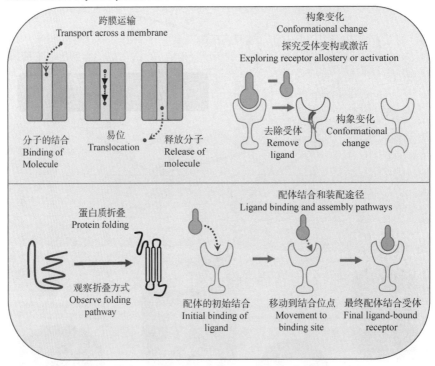

(a)

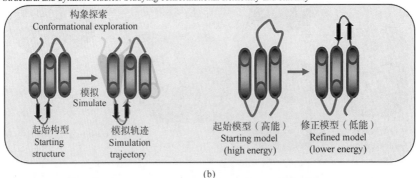

(b)

图 1-6-3 分子动力学模拟的一些普遍应用实例

Fig. 1-6-3 Some of the common applications of MD simulations

(a) 流程；(b) 结构和动力学研究

(a) Process；(b) Structural and dynamic studies

扰动：观察对系统施加可控更改后的响应
Perturbations: Observe response following controlled change to system

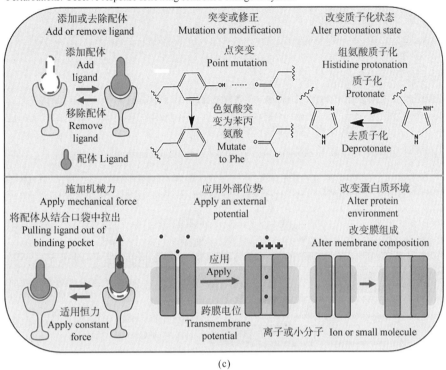

(c)

图 1-6-3　分子动力学模拟的一些普遍应用实例（续）

Fig. 1-6-3　Some of the common applications of MD simulations（continued）

（c）扰动

（c）Perturbations

Ⅶ 结构生物学
Ⅶ Structural Biology

7.1 结构生物学的研究对象及工具
7.1 Research Objects and Tools of Structural Biology

图 1-7-1 示出了结构生物学的研究对象及工具。

Fig. 1-7-1 shows research objects and tools of structural biology.

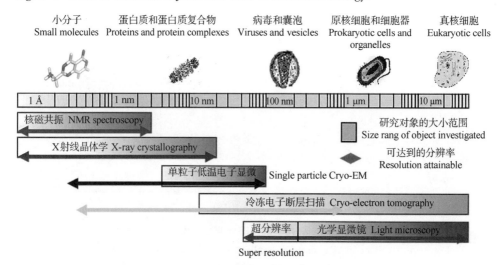

图 1-7-1 结构生物学技术及其研究对象（见彩插）

Fig. 1-7-1 Research objects and tools of structural biology (see the color figure)

可研究的生物对象的尺寸范围用粗标尺表示，而相应颜色的箭头表示可靶向的分辨率范围。

The size range of biological objects that can be studied is represented with thick bars, while corresponding color arrows indicate the resolution ranges that can be targeted.

7.2 揭示生物分子系统结构信息的方法
7.2 Methods for Revealing Structural Information of Biomolecular Systems

图 1-7-2 揭示了生物分子系统结构信息的方法。
Fig. 1-7-2 shows methods for revealing structural information of biomolecular systems.

```
                    化学计量学
                    Stoichiometry
                    • 质谱 • 荧光定量成像
                    • MS • Quantitative fluorescence imaging

                    结合位点映射
                    Binding site mapping
                    • 核磁共振 • 光谱学 • 诱变 • 交联质谱
                    • NMR • Spectroscopy • Mutagenesis, FRET • XL-MS
```

成对原子距离的大小、形状和分布	形状和尺寸
Size, shape, and distributions of pairwise atomic distances	Shape and size
• SAS 数据分析	• 原子力 • 显微镜离子迁移质谱
• SAS data analysis	• 荧光相关光谱 • 荧光各向异性 • 分析超速离心
	• Atomic force • Microscopy ion mobility mass spectrometry • Fluorescence correlation spectroscopy • Fluorescence anisotropy • Analytical ultracentrifugation

元件位置	物理相似度
Component positions	Physical proximity
• 超分辨率光学显微镜 • 荧光能量共振转移成像	• 共纯化 • 原生质谱 • 交联质谱 • 分子遗传学方法
• 免疫电子显微镜	• 基因/蛋白质序列协方差
• Super-resolution optical microscopy • FRET imaging • Immuno-electron microscopy	• Co-purification • Native mass spectrometry • XL-MS • Molecular genetic methods • Gene/protein sequence covariance

```
            不同基因组片段之间的相似度
            Proximity between different genome segments
            • 染色体构象捕获
            • Chromosome conformation capture
```

可溶性	不同交互模式的倾向分析
Solvent accessibility	Propensities for different interaction modes
• 印迹法 • 包括通过质谱或核磁共振评估的 HDex 点突变的功能结果	• 分子力学力场 • 平均力的潜力 • 统计潜力 • 序列共变
• Footprinting methods • including HDex assessed by MS or NMR functional consequences of point mutations	• Molecular mechanics force fields • Potentials of mean force • Statistical potentials • Sequence co-variation

原子距离和蛋白质距离	系统部分的原子结构
Atomic and protein distances	Atomic structures of parts of the studied system
• 核磁共振 • 荧光能量共振转移 • 其他荧光技术 • DEER, EPR • 其他光谱技术 • 交联质谱	• X 射线和中子 • 晶体学 • 核磁共振光谱, 3DEM • 同源建模 • 分子对接
• NMR • FRET • Other fluorescence techniques • DEER, EPR • Other spectroscopic techniques • XL-MS	• X ray and neutron • crystallography • NMR spectroscopy, 3DEM • Comparative modeling • Molecular docking

```
                3D 地图和 2D 图像
                • 电子显微镜和断层扫描
                3D maps and 2D images
                • Electron microscopy and tomography
```

图 1-7-2 生物分子系统结构信息的方法
Fig. 1-7-2 Methods for revealing structural information of biomolecular systems

7.3 蛋白质数据库
7.3 The Protein Data Bank

如图 1-7-3 所示，在过去的 30 年里，蛋白质数据库（PDB）中的条目几乎呈指数级增长。PDB 于 1971 年在布鲁克海文国家实验室成立时只有 7 个条目。条目总数在 1980 年增至 69 个，1990 年增至 507 个，2000 年增至 13 597 个，2010 年增至 70 039 个。截至 2019 年，PDB 共包含 154 478 个条目，其中 89.3% 的条目由 X 射线结晶测定，8.2% 通过核磁共振测定，2.3% 通过电镜测定。

As shown in Fig. 1-7-3, entries in the protein data bank (PDB) have enjoyed near-exponential growth in the past 30 years. PDB was established in 1971 at the Brookhaven National Laboratory with only seven entries. The total number of entries grew to 69 in 1980, 507 in 1990, 13,597 in 2000, and 70,039 in 2010. As of 2019, there were 154,478 total entries in PDB, of which 89.3% were determined by X-ray crystallography, 8.2% by NMR, and 2.3% by EM.

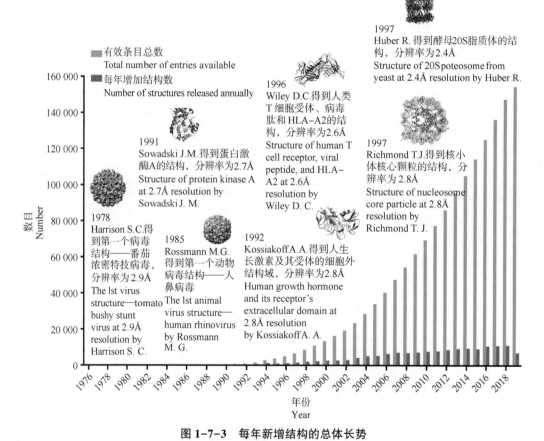

图 1-7-3　每年新增结构的总体长势

Fig. 1-7-3　Overall growth of released protein structures per year

图 1-7-4 所示为 PDB 中膜蛋白结构的总数。与膜蛋白的大量存在形成鲜明对比的是，PDB 中的膜蛋白结构只超过总条目的 1%，截至 2014 年 8 月 31 日，共有 1 520 个条目和

499 个结构。

Fig. 1-7-4 shows the total number of membrane protein structures in PDB. In sharp contrast to their heavy presence, structures of membrane proteins only account for just over 1% of all entries in the PDB, with 1,520 total entries and 499 unique structures as of August 31, 2014.

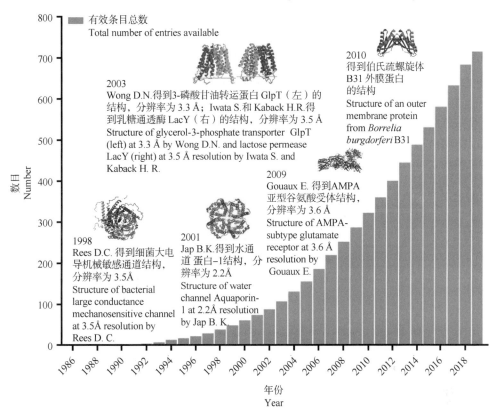

图 1-7-4 每年的膜蛋白结构的总数

Fig. 1-7-4 The total number of membrane protein structures in PDB

尽管近期取得了一些成功，高分辨率膜蛋白结构的获取仍存在很多瓶颈（图 1-7-5），包括表达量低、提取量低、纯化收率低，以及缺乏有序的三维晶体。然而，膜蛋白结构生物学领域正处于"对数增长"阶段。近年来，为克服膜蛋白 X 射线结构测定中存在的诸多障碍，人们进行了大量的研究工作并取得了长足进展，包括：①重组膜蛋白在不同宿主中的过表达；②新型洗涤剂和脂质的开发，以便更有效地增溶和结晶；③通过突变、缺失、融合伴侣工程和单克隆抗体手段提高蛋白质的稳定性，促使结晶质量达到衍射要求；④自动化、小型化和集成化的发展促进了初始结晶条件和晶体优化策略数量的增加；⑤同步辐射和束流的发展。

In spite of recent successes, the path to a high-resolution structure of a membrane protein still involves several bottlenecks (Fig. 1-7-5) including poor expression, low extraction, low purification yields and paucity of well-ordered 3D crystals. However, the field of membrane protein structural biology is in a "log" phase. In recent years, much effort has been put toward and innovative developments have been achived to overcome the numerous obstacles associated with X-ray structure determination of membrane proteins. For instance, much progress has been made regarding: ①overexpression of recombinant membrane proteins in different expression hosts; ②development of new detergents and lipids for more efficient solubilization and crystallization; ③improvement in protein stability through mutations, deletions, engineering of fusion partners and monoclonal antibodies, to promote diffraction quality crystals; ④development in automation, miniaturization and integration which have contributed to the increasing number of initial crystallization conditions and crystal optimization strategies; ⑤synchrotron radiation and beamline development.

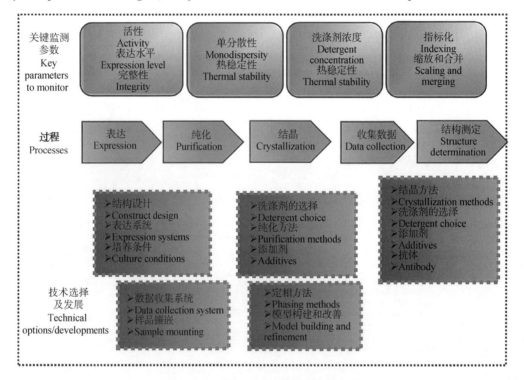

图 1-7-5 膜蛋白结构测定的流程和瓶颈

Fig. 1-7-5 Processes and bottlenecks in membrane protein structure determination

Ⅷ 生物传感器
Ⅷ Biosensors

8.1 生物传感器的定义与应用
8.1 Definition and Application of Biosensors

生物传感器是一种对生物物质敏感并能将其浓度转换为可检测电信号的仪器，是由固定化的生物敏感材料作为识别元件（包括酶、抗体、抗原、微生物、细胞、组织、核酸等其他生物活性物质）、适配的理化变换器（如氧电极、光敏管、场效应管、压电晶体等）及信号放大装置构成的分析工具或系统。生物传感器具有接收器与转换器的功能。图 1-8-1

所示为生物传感器的定义与应用。

Biosensor is an instrument that is sensitive to biological substances and converts its concentration into electrical signals for detection. It is made of immobilized biologically sensitive materials as identification elements (including enzymes, antibodies, antigens, microorganisms, cells, tissues, nucleic acids and other biologically active substances), appropriate physical and chemical transducers (such as oxygen electrodes, photosensitive tubes, field effect tubes, piezoelectric crystals, etc.) and signal amplification devices constitute analysis tools or systems. The biosensor has the functions of a receiver and a converter. Fig. 1-8-1 shows the definition and application of biosensors.

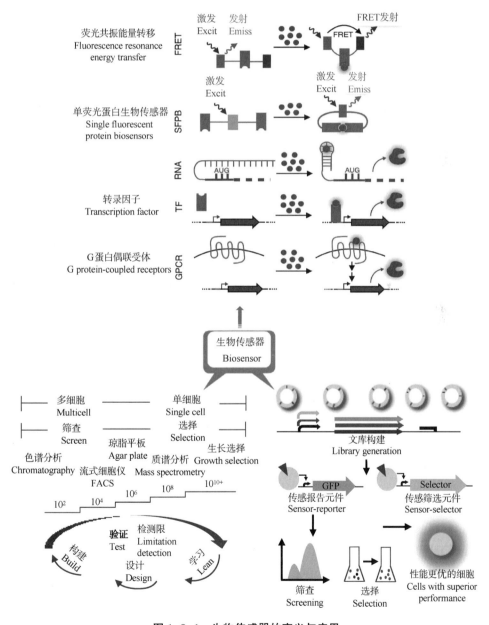

图 1-8-1　生物传感器的定义与应用

Fig. 1-8-1　Definition and application of biosensors

8.1.1　G 蛋白
8.1.1　G Proteins

G 蛋白，也称为鸟嘌呤核苷酸结合蛋白，是一个蛋白质家族，在细胞内充当分子开关，可将细胞外的各种刺激信号传递到细胞内。

G proteins, also known as guanine nucleotide-binding proteins, are a family of proteins that act as molecular switches inside cells, and are involved in transmitting signals from a variety of stimuli outside a cell to its interior.

8.1.2　基因文库
8.1.2　Genetic Library

基因文库包括基因组文库和部分基因文库。将含有某种生物不同基因的多个 DNA 片段导入受体菌群中储存，每个受体菌均含有这种生物的不同基因，称为基因文库。

Genetic library includes genome libraries and partial gene libraries. Many DNA fragments containing different genes of a certain organism are introduced into a population of recipient bacteria and stored. Each recipient bacteria containing different genes of this organism is called a gene library.

8.2　生物传感器在高通量筛选中的潜在应用
8.2　Potential Applications of Biosensors in High-throughput Screening

图 8-1-2 (a) 所示为以荧光灯信号作为输出的生物传感器，可用于筛选。目标代谢物（MOI，橙色）被生物传感器识别，驱动输出信号（绿色）的表达，信号的强度与目标代谢物浓度成比例。通常情况下可将荧光作为输出信号，用于在筛选（如流式细胞仪 FACS 或酶标仪检测）体系中分离高产突变体。图 1-8-2 (b) 所示为具有选择性输出的生物传感器。目标代谢物驱动某种蛋白（紫色）的表达，可赋予细胞生长优势。根据该图中和抗生素（红色）的酶以配体依赖性方式表达。在选择压力下（如存在抗生素），个体的生长差异将导致高产突变体在菌群中富集；图 1-8-2 (c) 所示为用于途径动态调控的生物传感器。该图呈现了通过抑制途径中的前一种酶并激活途径中的后一种酶来实现途径中间体的平衡。图 1-8-2 (d) 所示为途径平衡。管道表示酶的活性。酶浓度（通过转录因子调控或降解）或活性（变构效应）的调节可改变途径中间体之间的通量。当酶的通量失衡时（前两个例子），原料或中间体累积且产物合成受限。当酶的通量平衡时（后两个例子），各步骤均不会发生累积；图 1-8-2 (e) 所示为潜在的生物燃料合成途径和生物传感器。结合蛋白和转录因子（或受调节的启动子）以颜色来指示：蓝色表示还未用作生物传感器；紫色表示已被证明可用作生物传感器；绿色表示已用于筛查，选择或途径平衡的生物传感器。

Fig. 1-8-2 (a) shows a biosensor with an output, such as fluorescence, can be used in a screen. The metabolite of interest (MOI, orange) is detected by a biosensor, which drives the expression of an output signal (green) in proportion to the MOI concentration. The output signal, often fluorescence, is used to isolate high-producing variants through screening (e. g. , FACS or plate reader assays). Fig. 1-8-2 (b) shows a biosensor with a selectable output. The MOI drives the expression of a protein (purple) that provides a growth advantage to the cell. As depicted, an enzyme that neutralizes an antibiotic (red) is expressed in a ligand-dependent manner.

Growing variants under selective pressure (e.g., in the presence of antibiotic) enriches the population with high-producing variants; Fig. 1-8-2 (c) shows biosensors for dynamic regulation of pathways. The MOI (orange) is detected by a biosensor (biosensor actions are represented by feedback symbols) which alters the expression of enzymes in the MOI pathway. The cartoon illustrates the balancing of a pathway intermediate (MOI, orange) by repressing the preceding enzyme and activating the subsequent enzyme in the pathway. Fig. 1-8-2 (d) shows visualization of pathway balancing. Enzyme activity is represented by tubes. Regulation of enzyme concentration (by TFs or degradation) or activity (allostery) alters the flux capacity between pathway intermediates. When enzyme flux is imbalanced (top two examples), starting materials or intermediates accumulate and product formation is limited. When flux is balanced (bottom two examples), accumulation does not occur at any step. Fig. 1-8-2 (e) shows potential biofuel pathways and biosensors. Binding proteins and transcription factors (or regulated promoters) are shown with colors designating: known binder with no biosensor use published (blue). Demonstrated use as a biosensor (purple), or biosensor applied to screening, selection, or pathway-balancing (green).

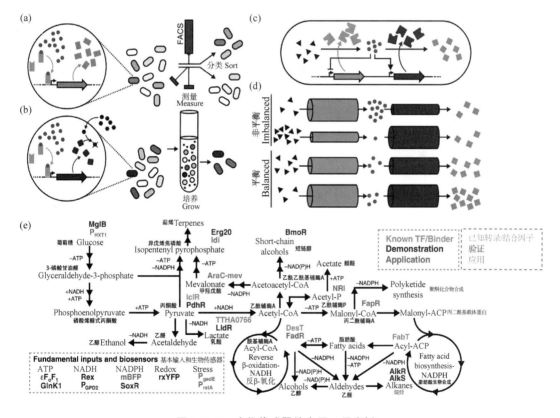

图 1-8-2 生物传感器的应用（见彩插）

Fig. 1-8-2 Applications of biosensors (see the color figure)

(a) 一个以荧光灯信号作为输出的生物传感器；(b) 具有选择性输出的生物传感器；
(c) 用于途径动态调控的生物传感器；(d) 途径平衡；(e) 潜在的生物燃料合成途径和生物传感器

(a) A biosensor with an output; (b) A biosensor with a selectable output; (c) Biosensors for dynamic regulation of pathways; (d) Visualization of pathway balancing; (e) Potential biofuel pathways and biosensors

Ⅸ 高通量筛选技术
Ⅸ High-throughput Screening Technology

9.1 高通量筛选方案
9.1 High-throughput Screening Strategies

为建立筛选高性能微生物的高通量筛选体系，已开发了凸显目标产物的不同策略方案。图 1-9-1（a）所示为基于吸光度的策略，包括①基于目标产品的分子结构或自身颜色特性的直接检测，②间接检测，如通过添加 pH 指示剂、金属离子螯合以及与酶促或化学反应偶联；图 1-9-1（b）所示为基于荧光的策略，包括①基于目标产物的自身荧光进行直接检测，②使用染料/pH 探针、金属离子、与酶促或化学反应偶联的间接检测；图 1-9-1（c）所示为基于生物传感器的策略，包括①转录因子（TF），②荧光蛋白（FPs）和 Förster 共振能量转移（FRET）检测，③基于核糖体结合位点（RBS）的 RNA 核糖开关，④基于核酶的 RNA 核糖开关，⑤Spinach RNA 适配体，⑥DNA 生物传感器；图 1-9-1（d）所示为基于电化学传感器（ES）的策略；图 1-9-1（e）所示为基于特殊光谱的策略，如拉曼（RS）、傅里叶变换红外（FTIR）和傅里叶变换近红外（FTNIR）光谱。

Different strategies have been employed to highlight the target products for establishing high-throughput screening for screening high-performance microorganisms. Fig. 1-9-1（a）shows absorbance-based strategies which include ①direct detection based on the molecular structure or an innate color property of the target product, ②indirect detection by, for example, the addition of a pH indicator, metal ion chelations, and coupling with enzymatic or chemical reactions; Fig. 1-9-1（b）shows fluorescence-based strategies which include ①direct detection based on innate fluorescence of the target product, ②indirect detection using a dye/pH probe, metal ions, and coupling with enzymatic or chemical reaction; Fig. 1-9-1（c）shows biosensor-based strategies which include ①transcription factors（TFs）, ②fluorescent proteins（FPs）and detection by Förster resonance energy transfer（FRET）, ③RNA riboswitch based on ribosome binding sites（RBS）, ④RNA riboswitch based on ribozymes, ⑤spinach RNA aptamer, ⑥DNA biosensors; Fig. 1-9-1（d）shows electrochemical sensor（ES）-based strategies; Fig. 1-9-1（e）shows special spectrum-based strategies such as Raman（RS）, Fourier transform infrared spectroscopy（FTIR）, and Fourier transform near-infrared spectroscopy（FTNIR）spectroscopy.

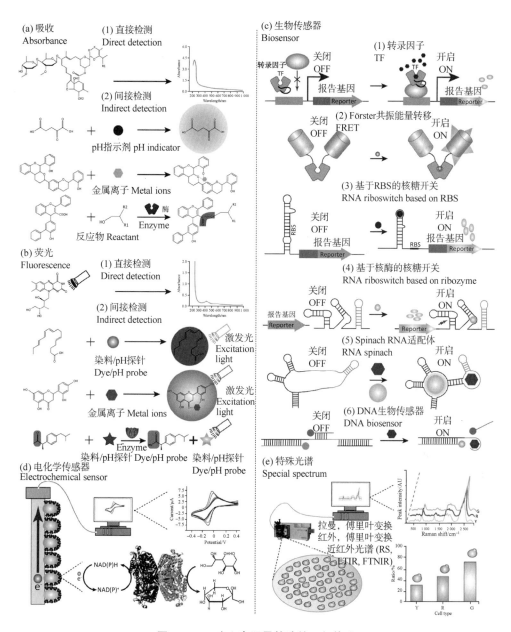

图 1-9-1 建立高通量筛选的一般策略

Fig. 1-9-1 General strategies for establishing high-throughput screening

(a) 基于荧光的策略;(b) 基于荧光的策略;(c) 基于电化学传感器的策略;
(d) 基于生物传感器的策略;(e) 基于特殊光谱的策略

(a) Absorbance-based strategies; (b) Fluorescence-based strategies; (c) Biosensor-based strategies;
(d) Electrochemical sensor-based strategies; (e) Special spectrum-based strategies

高通量筛选是基于分子和细胞水平的实验方法,以微孔板作为实验工具,通过自动化操作系统执行测试过程,通过灵敏、快速的检测仪器采集实验结果数据,并通过计算机同时对实验数据进行分析和处理。该技术可检测数千万个样本并支持对所获得的数据库进行操作，具有溯源、快速、灵敏、准确的特点。总之，通过一次实验可以获得大量信息，并可从中发现有价值的信息。

Based on experimental methods at the molecular and cellular levels, microplates are used as the experimental tool carrier, the test process is executed with an automated operating system, the experimental result data is collected with sensitive and fast detection instruments, and the experimental data is analyzed and processed by a computer at the same time. The technical system that detects tens of millions of samples and supports operation with the corresponding database obtained, it has the characteristics of trace, fast, sensitive and accurate. In short, a large amount of information can be obtained through one experiment, and valuable information can be found from it.

9.2 基于FACS的高通量筛选
9.2 FACS-based High-throughput Screening

如图1-9-2（a）所示，选择携带荧光报告基因的转基因植物进行实验，该报告基因的表达具有细胞或组织特异性；如图1-9-2（b）所示，收获了样品并用酶处理使细胞解离，然后过滤以分解大细胞团块；如图1-9-2（c）所示，在典型的流式细胞仪中，细胞样本被注入鞘液分类流中，通过高频振动，将其破碎成均匀的液滴，每个液滴含有不超过一个细胞；如图1-9-2（d）所示，联合激光和探测仪来分析发射光谱并确定合适的波长"门"以定义用于分选的阳性液滴和阴性液滴；如图1-9-2（e）所示，分析发射光谱并确定适当的波长"门"，据此定义用于分选的正、负液滴；如图1-9-2（f）所示，每个门内的液滴都带有电荷；如图1-9-2（g）所示，电板偏转带电液滴，此处图示了对所有其他细胞（蓝色）进行废物收集的双向分选（绿色/灰色阴影细胞）；如图1-9-2（h）所示，将细胞收集到试管中然后用于观察分析（例如，荧光水平的确认）或收集到含有缓冲液的试管中以立即裂解细胞来进行快速分子提取。

As shown in Fig. 1-9-2 (a), transgenic plants carrying a fluorescent reporter conferring cell or tissue-specific expression are chosen for an experiment; As shown in Fig. 1-9-2 (b), samples are harvested and treated with enzymes to dissociate cells, which are then filtered to break up large cell clumps; As shown in Fig. 1-9-2 (c), in a typical FACS machine, the cell sample is injected into a sheath fluid sort stream, then the stream is vibrated at a high frequency to break it into uniform droplets containing no more than one cell each; As shown in Fig. 1-9-2 (d), laser and detectors are combined to measure the fluorescence and other properties (such as size) of each droplet; As shown in Fig. 1-9-2 (e), the emission spectrum is analyzed and appropriate "gates" of wavelength determined to define the positive and negative droplets for sorting; As shown in Fig. 1-9-2 (f), An electrical charge is imparted on droplets within each gate; As shown in Fig. 1-9-2 (g), electrical plates deflect charged droplets, here illustrated for two-way sorting (green/grey shaded cells) with a waste collection of all other cells (blue); As shown in Fig. 1-9-2 (h), cells are collected into tubes and then used for visual analysis (for instance, confirmation of fluorescence levels) or collected into tubes that contain buffer to immediately lyse cells for rapid molecular extraction.

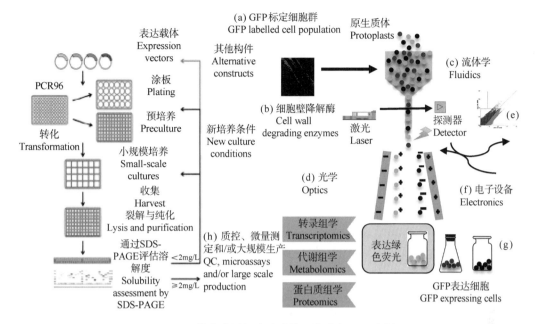

图 1-9-2 荧光激活细胞分选的工作流程（见彩插）

Fig. 1-9-2 Fluorescence-activated cell sorting workflow（see the color figure）

（a）选择携带荧光报告基因的转基因植物；（b）用酶处理收获样品并过滤；
（c）在流式细胞仪中处理样本；（d）联合激光和探测仪分析以定义液滴；（e）分析发射光谱；
（f）带有电荷的液滴；（g）所有其他细胞进行废物收集的双向分选；（h）收集细胞并分析

（a）Choose transgenic plants carrying a fluorescent reporter；（b）Sampling harvesting, treating with enzymes and filtering；
（c）Dealing with the sample in a FACS machine；（d）Measure the fluorescence and other properties of each droplet；
（e）Analyzing the emission spectrum；（f）Droplets Charged；
（g）Waste collection of all other cells；（h）Cells collecting and analyzing

X 酶固定化
X Enzyme Immobilization

图 1-10-1 所示为酶固定化方法及其应用。

Fig. 1-10-1 shows methods and applications of enzyme immobilization.

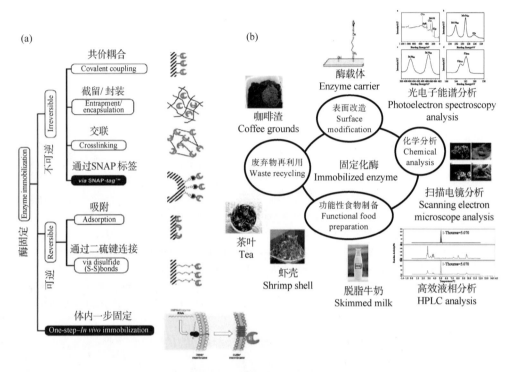

图 1-10-1 酶固定化的方法及应用

Fig. 1-10-1 Methods and applications of enzyme immobilization

（a）酶固定化的不同方法；（b）固定化酶的应用

(a) Different methods to immobilize enzymes; (b) Applications of immobilized enzymes

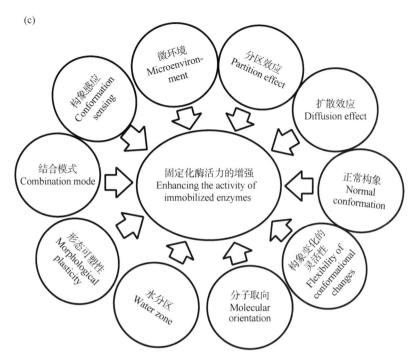

图 1-10-1 酶固定化的方法及应用（续）

Fig. 1-10-1 Methods and applications of enzyme immobilization (continued)

（c）影响固定化酶活性的因素

(c) Factors affecting the activity of immobilized enzymes

XI 无细胞催化
XI Cell-free Catalysis

11.1 无细胞催化技术及其关键流程
11.1 Cell-free Catalytic Technology and Key Processes

图1-11-1（a）所示为以混合组分诱导无细胞蛋白合成反应（CFPS）；图1-11-1（b）所示为有选择性地添加激活效应物（中间）以提高难表达蛋白的产量。无细胞蛋白合成反应利用来自原核、植物和哺乳动物细胞粗提物中的转录/翻译机器，或重组纯化的转录/翻译组件。裂解细胞，然后通过多轮洗涤和离心去除细胞碎片和大分子（如基因组DNA）来制备粗提取物。模拟细胞质环境是启动和增强无细胞蛋白合成的关键。因此，提取物中要加入能量、辅因子、盐、核苷酸和氨基酸等。加入环状质粒或线性PCR片段形式的DNA模板并在适当温度下孵育后，体外转录和翻译反应将同时启动，开始蛋白质的合成。蛋白质在数小时内合成，产生g/mL规模的批量生产；然而，通过向连续流动无细胞系统中补加含有核苷酸构建模块的缓冲液，以及使用集成透析膜的连续交换无细胞（CECF）系统可延长反应寿命（多至数天），并实现更高的蛋白质产量（多达数mg/mL）。

As shows in Fig. 1-11-1 (a), cell protein synthesis system (CFPS) reactions are induced by mixing components. Positive effectors are selectively added (middle) to improve the production of difficult-to-express proteins, as shown in Fig. 1-11-1 (b). CFPS utilizes transcriptional/translational machineries from crude cell extracts derived from prokaryotic, plant, and mammalian cells or reconstituted purified transcriptional/translational components. Crude extracts are prepared by lysing the cells followed by removing cellular debris and large molecules, such as genomic DNA, via multiple rounds of washing and centrifugation. It is critical to mimic the cytoplasmic environment to enable and enhance CFPS. Hence, energy sources, cofactors, salts, nucleotides, amino acids, etc., are mixed together with the extract. After adding DNA gene template in either circular plasmid or linear PCR fragment forms and incubating at a proper temperature, in vitro transcription and translation reactions occur simultaneously, resulting in protein production. Proteins are synthesized within several hours, yielding g/mL scale batch reaction; however, continuous-flow cell-free systems via feeding buffers containing nucleotide building blocks and continuous-exchange cell-free (CECF) settings using an integrated dialysis membrane allow a prolonged reaction lifetime (up to several days) and a higher protein yield (up to several mg/mL).

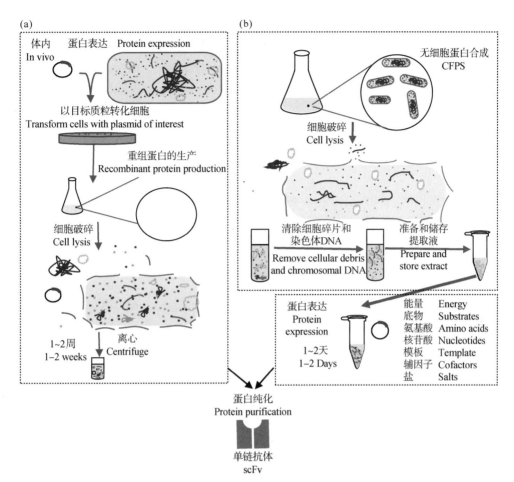

图 1−11−1　通过无细胞蛋白合成生产难表达蛋白

Fig. 1−11−1　Use of cell-free protein synthesis to produce difficult-to-express proteins

（a）以混合组分诱导无细胞蛋白合成反应；（b）选择性添加激活效应物

(a) CFPS reactions are induced by mixing components; (b) Positive effectors are selectively added

图 1−11−2 所示为体内重组 DNA 蛋白表达与无细胞蛋白合成的比较。

Fig. 1−11−2 shows comparison of *in vivo* recombinant DNA protein expression with cell-free protein synthesis.

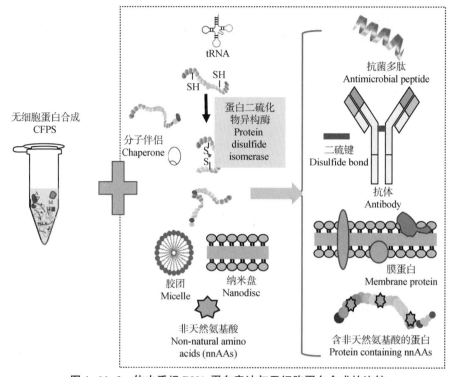

图 1-11-2 体内重组 DNA 蛋白表达与无细胞蛋白合成的比较

Fig. 1-11-2 Comparison of *in vivo* recombinant DNA protein expression with cell-free protein synthesis

11.2 无细胞系统来源、优化及应用
11.2 CFS Source, Optimization and Application

如图 1-11-3（a）所示，无细胞系统可以通过裂解大肠杆菌、小麦胚芽、利什曼原虫或海拉细胞产生。裂解物含有体外翻译所需的所有组件（核糖体和翻译装置，TraM）和其他蛋白质杂质。用于蛋白质合成的重组元件（PURE）系统则是通过纯化 TraM 和核糖体的各个部件来获得的。

如图 1-11-3（b）所示，来自 T7 和 T3 噬菌体的正交 RNA 聚合酶（RNAP）可用于无细胞系统中蛋白质的转录。此外，σ 转录因子可以指导 RNAP Ⅱ 从特定启动子起始转录。添加 DNA 外切酶（ExoDNAse）抑制剂 GamS 可以稳定线性 DNA。此外，引入伴侣或翻译后修饰因子（Post-Trn）也可提高蛋白质的稳定性。mRNA 和蛋白质的周转可通过引入特定的 RNA 内切酶（EndoRNAse）和/或识别目标蛋白中特定标签的蛋白酶来实现。还可通过正交和突变合酶向靶标蛋白中掺入新的非天然氨基酸，来生产具有扩展特性的靶标蛋白。

如图 1-11-3（c）所示，使用无细胞系统可进化出编码成孔蛋白的 DNA 文库。如果形成的孔隙合适，脂质体外环境中的荧光团（绿色星形）就会被转运到脂质体中。这样，通过流式细胞仪分选即可筛选出具有活性成孔突变体的脂质体。回收所选脂质体中的 DNA 并用于候选轮次的筛选、修改和诊断。可将无细胞系统连同响应特定分子的基因线路（传感器）印在纸上，添加来自患者的样本以诊断疾病生物标志物。

As shown in Fig. 1-11-3 (a), CFS can be generated through lysis of *E. coli*, wheat germ, *L. tarentolae* or HeLa cells. The lysates contain all the components required for *in vitro* translation (ribosomes and translation machinery, TraM) and other protein impurities. The protein synthesis using recombinant elements (PURE) system is obtained by purifying each of the components of the TraM and the ribosomes.

As shown in Fig. 1-11-3 (b), orthogonal RNA polymerases (RNAP) from phage T7 and T3 can be used to transcribe proteins in CFS. In addition, σ transcription factors can guide the transcription from particular promoters by RNAP Ⅱ. Stabilization of linear DNA can be achieved by adding DNA exonuclease (ExoDNAse)-inhibitor GamS. Furthermore, protein stability can be increased using chaperones or factors that introduce post-translational (Post-Trn) modifications. Turnover of mRNA and proteins can be achieved by incorporating specific RNA endonucleases (EndoRNAse) and/or proteases that recognize specific tags in the target protein. Target proteins with extended properties can also be produced by incorporating novel, non-natural amino acids through orthogonal and mutant synthases.

As shown in Fig. 1-11-3 (c), DNA library coding for a pore forming protein is evolved using CFS. A fluorophore (green stars) in the extraliposomal milieu is transported into the liposome if the pore is properly formed. This way, liposomes with active pore-forming mutants are selected by flow cytometry sorting. The DNA in the selected liposomes is recovered and used in successive rounds of selection, modification and diagnosis. The CFS, together with a genetic circuit (sensor) responsive to specific molecules are printed on a piece of paper. Samples from patients can be added to the paper for diagnosis of disease biomarkers.

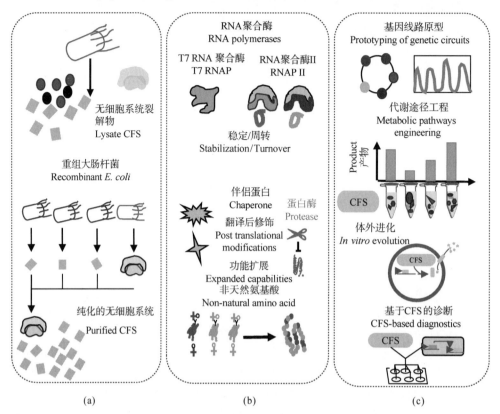

图 1-11-3 无细胞系统来源、优化及应用（见彩插）

Fig. 1-11-3 CFS source, optimization and application (see the color figure)

(a) 合成生物学方法制备 CFS；(b) CFS 的优化和扩展；(c) CFS 在合成生物学中的应用

(a) CFS for synthetic biology approaches；(b) Optimization and expansion of CFS；

(c) Applications for CFS in synthetic biology

11.2.1 纳米盘
11.2.1 Nanodisc

纳米盘是一种合成的模式膜系统，有助于膜蛋白的研究。它由磷脂的脂质双分子层组成，疏水边缘被两种两亲性蛋白质掩蔽。这些蛋白质被称为膜支架蛋白（MSP），并以双带形式排列。纳米盘在结构上与高密度脂蛋白（HDL）非常相似，而 MSP 是 HDL 的主要成分载脂蛋白 A1（apoA1）的修饰版本。由于可以溶解和稳定膜蛋白，且比脂质体、洗涤剂胶束、双层膜微胞和两性分子更能代表原生环境，纳米盘适用于膜蛋白的研究。

A nanodisc is a synthetic model membrane system which assists in the study of membrane proteins. It is composed of a lipid bilayer of phospholipids with the hydrophobic edge screened by two amphipathic proteins. These proteins are called membrane scaffolding proteins (MSP) and align in double belt formation. Nanodiscs are structurally very similar to high-density lipoproteins (HDL) and the MSPs are modified versions of apolipoprotein A1 (apoA1), the main constituent in HDL. Nanodiscs are useful in the study of membrane proteins because they can solubilize and stabilize membrane proteins and represent a more native environment than liposomes, detergent micelles, bicelles and amphipols.

11.2.2 非天然氨基酸
11.2.2 Non-natural amino acids

非天然氨基酸可用于扩展蛋白质序列空间,使其显著超出自然设定的限制。扩展蛋白质序列空间为改造和化学修饰蛋白质打开了一扇新的大门。需要将密码子重新分配给非天然氨基酸并对蛋白翻译机器进行改造,以将非天然氨基酸整合到细胞中靶蛋白的单个或多个位点。

Non-natural amino acid can be used to expand the protein sequence space significantly beyond the limits set by nature. Expanding the protein sequence space opens a new door to engineering and chemically modifying proteins. Reassigning codons to non-natural amino acids as well as engineering protein translational machinery is required to incorporate non-natural amino acids into a single or multiple sites of a target protein in cells.

XII 人工混菌体系
XII Synthetic Microbial Consortia

分别用纯培养或混合菌体系将底物(S)转化为产物(P)。在纯培养中,需要对代谢反应进行区室化以转化胞内的有毒代谢物(M)。在混合培养中,直接的细胞-细胞相互作用和中间体依赖的相互作用(底物/产物抑制)都参与稳定菌群的形成。

图1-12-1所示为由两种微生物组成的混菌体系的设计、分析与应用。图1-12-1(a)所示为 *G. metallireducens* 与 *M. harundinacea* 之间的电子传递;图1-12-1(b)所示为 *B. megaterium* 与 *K. vulgare* 之间的代谢物交换;图1-12-1(c)将氧合紫杉烷代谢途径在 *E. coli* 与 *S. cerevisiae* 混菌体系内分配;图1-12-1(d)所示为未来的细菌治疗体系。

Conversion of substrates (S) to product (P) using pure or mixed microbial cultures respectively. Compartmentalization of metabolicreaction is needed to convert some toxic metabolites (M) within the cells in a pure culture. In a mixed culture, both direct cell-cell interaction and intermediates dependent interaction (substrate/product inhibition) are involved for the development of a stable consortium.

Fig. 1-12-1 shows design, analysis and application of synthetic microbial consortia composed of two species. Fig. 1-12-1 (a) shows electron transfer between *G. metallireducens* and *M. harundinacea*; Fig. 1-12-1 (b) shows metabolites exchange between *B. megaterium* and *K. vulgare*; Fig. 1-12-1 (c) shows distribute oxygenated taxanes metabolic pathway into *E. coli* and *S. cerevisiae* consortium; Fig. 1-12-1 (d) shows bacterial cell therapy systems in the future.

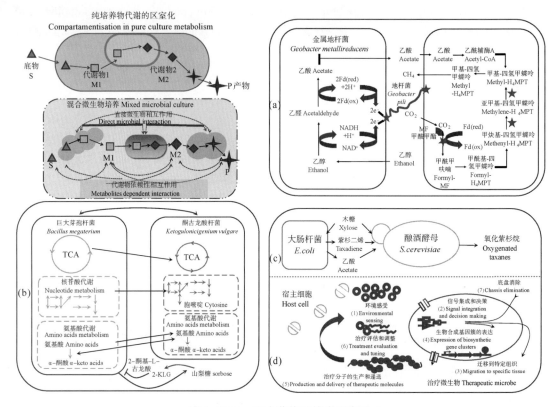

图 1-12-1 混合菌体系的应用案例

Fig. 1-12-1 Application of synthetic microbial consortia

(a) *G. metallireducens* 与 *M. harundinacea* 之间的电子传递；

(b) *B. megaterium* 与 *K. vulgare* 之间的代谢物交换；

(c) *E. coli* 与 *S. cerevisiae* 混菌体系内分配； (d) 未来的细菌治疗体系

(a) Electron transfer between *G. metallireducens* and *M. harundinacea*；

(b) Metabolites exchange between *B. megaterium* and *K. vulgare*；

(c) Distribute oxygenated taxanes metabolic pathway into *E. coli* and *S. cerevisiae* consortium；

(d) Bacterial cell therapy systems in the future

第 2 章　工业生物技术

Chapter 2　Industrial Biotechnology

本章提要

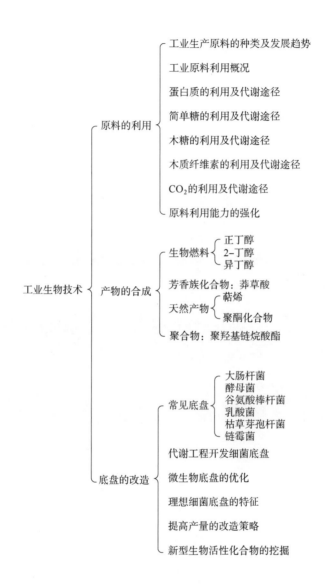

Summary of Chapter 2

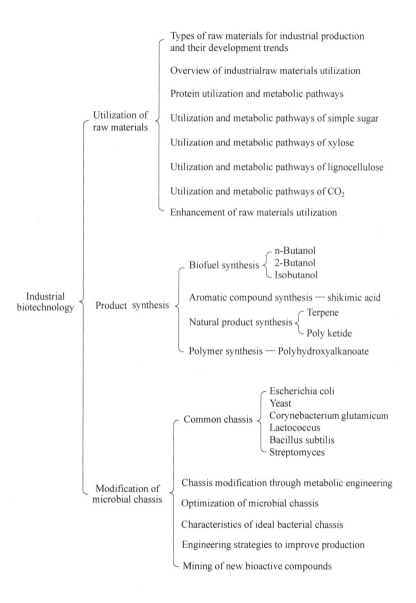

1 原料的利用
1 Utilization of Raw Materials

1.1 工业生产原料的种类及发展趋势
1.1 Types of Raw Materials for Industrial Production and Their Development Trends

图 2-1-1 所示为工业生产原料的种类及其发展趋势。图 2-1-1（a）所示为循环生物经济；图 2-1-1（b）所示为生物燃料生产原料的演变过程；图 2-1-1（c）所示为生物质潜力的主要组成；图 2-1-1（d）所示为根据生物质政策估算德国 2010 年、2020 年、2030 年不同门类和原料类型的生物质潜力。

Fig. 2-1-1 shows types of raw materials for industrial production and development trends; Fig. 2-1-1 (a) shows circular Bioeconomy; Fig. 2-1-1 (b) shows evolution of biofuel production feedstock; Fig. 2-1-1 (c) shows major components of biomass potential; Fig. 2-1-1 (d) shows biomass policies estimated biomass potentials for Germany for 2010, 2020, 2030 under different sector and feedstock types.

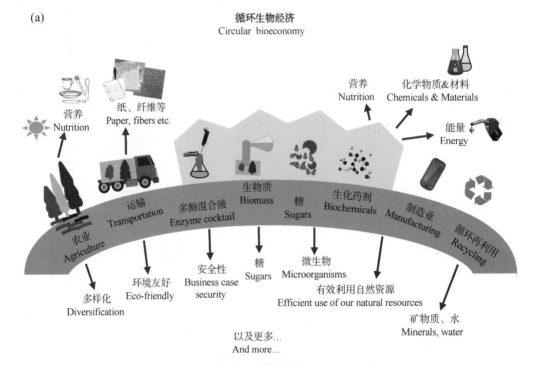

图 2-1-1 工业生产原料的种类及发展趋势

Fig. 2-1-1 Types of raw materials for industrial production and their development trends

(a) 循环生物经济

(a) Circular Bioeconomy

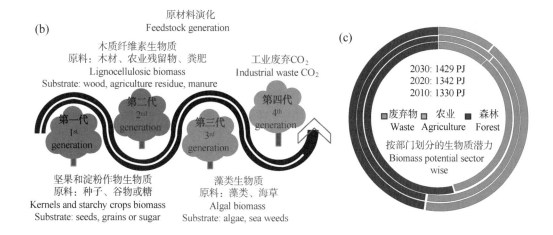

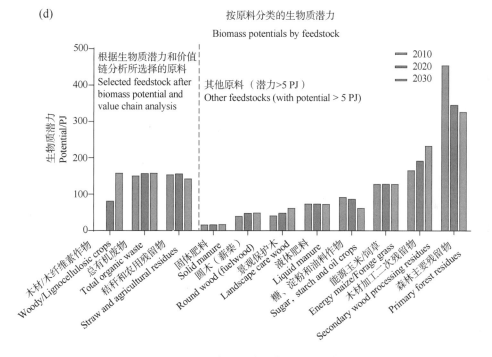

图 2-1-1 工业生产原料的种类及发展趋势（续）

Fig. 2-1-1 Types of raw materials for industrial production and their development trends (continued)

(b) 生物燃料生产原料的演变过程；(c) 生物质潜力的主要组成；

(d) 根据生物质政策估算德国 2010 年、2020 年、2030 年不同门类和原料类型的生物质潜力

(b) Evolution of biofuel production feedstock；(c) Major components of biomass potential；

(d) Biomass policies estimated biomass potentials for Germany for 2010, 2020, 2030 under different sector and feedstock types

注释：

PJ 是一种能量单位，相当于约 3 000 万 kW·h。

Note：

PJ is a measure of energy consumption, equivalent to approximately 30 million kW·h.

1.2 工业原料利用概况
1.2 Overview of Industrial Raw Materials Utilization

整个生物质价值链都有巨大的收益潜力（图2-1-2），到2020年，商业潜力约为数十亿美元。

There are significant revenue potentials along the entire biomass value chain（Fig. 2-1-2）. The values given are approximate billions of U.S. dollars by 2020.

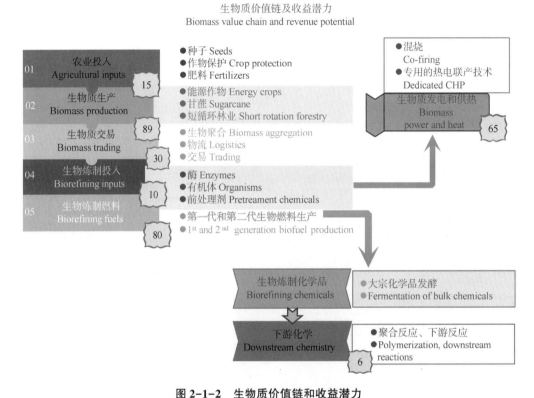

图2-1-2 生物质价值链和收益潜力

Fig. 2-1-2 Biomass value chain and revenue potential

图2-1-3为第一、第二代生物质原料的主要生物燃料的合成途径。

In Fig. 2-1-3 shows the synthetic conversion routes of major biofuels produced from the 1st and 2nd generation biomass feedstock.

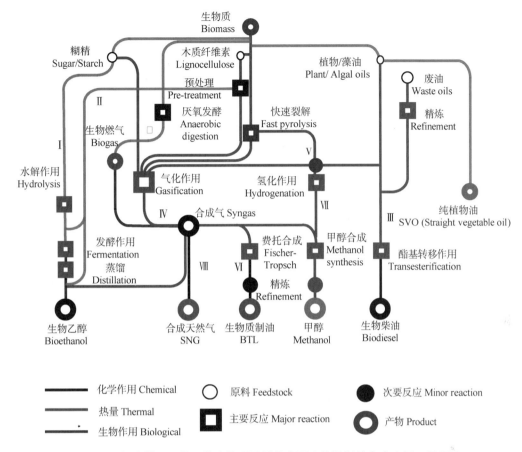

图 2-1-3 源自第一、第二代生物质原料的主要生物燃料的合成途径（见彩插）
Fig. 2-1-3 The synthetic conversion routes of major biofuels produced from the 1st and 2nd generation biomass feedstock (see the color figure)

注释：
Ⅰ，糖/淀粉作物的发酵；Ⅱ，木质纤维素生物质的发酵；Ⅲ，甘油三酯的酯交换；Ⅳ，合成气；Ⅴ，快速裂解；Ⅵ，费托合成；Ⅶ，氢化；Ⅷ，合成天然气。
Note：
Ⅰ, Fermentation of sugar/starch crops; Ⅱ, Fermentation of lignocellosic biomass; Ⅲ, Transesterification of triglycerides; Ⅳ, Syngas; Ⅴ, Fast pyrolysis; Ⅵ, Fischer-Tropsch synthesis; Ⅶ, Hydrogenation; Ⅷ, Synthetic natural gas (SNG).

转化技术：根据原料和目标产物，生物精炼厂采用各种转化技术，将原料生物质转化为商用能源。这些技术通常包括发酵、气化和酯交换。在合成生物燃料如生物质制油（BTL）的开发中，新的非传统方法受到了持续关注。目前还可以将生物质转化为新型化学品和材料，但其商业开发的程度还远低于燃料。

Conversion techniques: based on the feedstock and the desired output, biorefineries employ a variety of conversion technologies, transforming the raw biomass into commercial energy sources. These technologies typically include fermentation, gasification and transesterification. New non-traditional methods are constantly being investigated, particular in the development of synthetic biofuels such as biomass-to-liquid (BTL). Novel chemicals and materials produced from biomass are also currently available, but it is still far less commercially developed than fuels.

生物质制油：将生物质通过热化学途径转化成液体生物燃料的多步骤过程。它利用整个工厂来平衡二氧化碳并提高燃料产量，这种燃料被称为草料油。

Biomass to liquid (BTL or BMTL): multi-step process in which biomass is thermochemically produced into liquid biofuel and utilizes the whole factory to balance carbon dioxide and increase the production of fuel, known as grassoline.

费托合成：又称F-T合成，是以合成气（一氧化碳和氢气的混合气体）为原料在催化剂和适当条件下合成液态的烃或碳氢化合物（hydrocarbon）的工艺过程。费托工艺包括一系列生成多种烃类的化学反应，其中生产烷烃的用途较广，其反应方程式如下：

$$(2n+1)H_2 + nCO \rightarrow C_nH_{(2n+2)} + nH_2O$$

式中，$C_nH_{(2n+2)}$为烷烃的通式。

Fischer-Tropsch process: Also known as F-T synthesis, it is a process in which syngas (carbon monoxide and hydrogen gas mixture) is used as raw material to synthesize liquid hydrocarbon or hydrocarbon under catalyst and appropriate conditions. The Fischer-Tropsch process consists of a series of chemical reactions to generate a variety of hydrocarbons, among which the production of alkanes is more widely used. The reaction equation is shown below:

$$(2n+1)H_2 + nCO \rightarrow C_nH_{(2n+2)} + nH_2O$$

where alkanes are represented by the general formula $C_nH_{(2n+2)}$.

代谢工程已经开发出可以将可再生碳源转化为生物燃料的微生物细胞工厂。目前的分子生物学工具可以有效地改变酶的水平从而将碳流重新定向到生物燃料生产，但是大型生物反应器中的低产物产率和产量阻止了廉价生物燃料的获得。阻碍生物燃料生产经济性的主要障碍有三个。首先，来自底物的碳流消散入复杂的代谢网络。其次，微生物宿主需要氧化大部分底物以产生ATP和NAD(P)H，为生物燃料合成提供动力。最后，大型生物反应器中的传质限制造成了异质生长条件和微环境波动（例如次优O_2水平和pH值），导致代谢压力和遗传不稳定性。为了克服这些限制，发酵工程应该与系统代谢工程相结合。现代发酵工程需要采用新的代谢流分析工具，整合动力学，流体动力学和^{13}C蛋白质组学，以揭示在大型生物反应器条件下微生物宿主的动态生理特征。基于代谢分析，发酵工程可以采用理性途径改造、合成生物学回路，以及生物反应器控制算法来优化大规模生物燃料的生产。图2-1-4所示为生物合成的碳和能量限制。

Metabolic engineering has developed microbial cell factories that can convert renewable carbon sources into biofuels. Current molecular biology tools can efficiently alter enzyme levels to redirect carbon fluxes toward biofuel production, but low product yield and titer in large bioreactors prevent the fulfillment of cheap biofuels. There are three major roadblocks preventing economical biofuel production. First, carbon fluxes from the substrate dissipate into a complex metabolic network. Second, microbial hosts need to oxidize a large portion of the substrate to generate both ATP and NAD (P) H to power biofuel synthesis. Third, mass transfer limitations in large bioreactors create heterogeneous growth conditions and micro-environmental fluctuations (such as suboptimal O_2 level and pH value) that induce metabolic stresses and genetic instability. To overcome these limitations, fermentation engineering should merge with systems metabolic engineering. Modern fermentation engineers need to adopt new metabolic flux analysis tools that integrate kinetics, hydrodynamics, and ^{13}C-proteomics, to reveal the dynamic physiologies of the microbial host under large bioreactor conditions. Based on metabolic analyses, fermentation engineers may employ rational pathway modifications, synthetic biology circuits, and bioreactor control algorithms to optimize large-scale biofuel production. Fig. 2-1-4 shows carbon and energy limitation for biosynthesis.

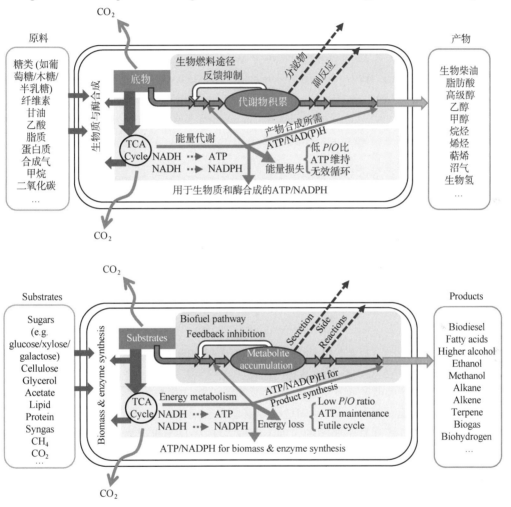

图 2-1-4 生物合成的碳和能量限制
Fig. 2-1-4 Carbon and energy limitation for biosynthesis

生物燃料途径：除了利用自然界的现有生产途径之外，研究人员还可将已知基因组装成非天然途径，来生产针对特定应用而优化的燃料。可再生生物燃料途径包括三个关键组分：原料、生产过程、燃料类型。

Biofuels pathways: In addition to exploiting existing production pathways in nature, researchers have created new pathways by incorporating known gene combinations into non-natural pathways to produce fuels optimized for specific applications. The renewable biofuel pathway consists of three key components: feedstock, production process, fuel type.

磷氧比：电子经过呼吸链的传递作用最终与氧结合生成水，在此过程中所释放的能量用于 ADP 磷酸化生成 ATP。此过程中，消耗一个氧原子所需要无机磷酸的分子数（也是生成 ATP 的分子数）称为磷氧比（P/O）。NADH 的磷氧比是 3，$FADH_2$ 的磷氧比是 2。P/O 取决于通过电化学梯度向外输送的氢原子数，以及通过诸如 ATP 合酶等向内跨膜返回的质子数。

P/O ratio: Electrons eventually combine with oxygen to form water through the transfer of respiratory chain, and the energy released in this process is used for phosphorylation of ADP to generate ATP. The number of molecules of inorganic phosphoric acid (also the number of molecules that produce ATP) that it takes to consume an atom of oxygen in this process is called the phosphorus/oxygen ratio (P/O). NADH has a P/O ratio of 3, and $FADH_2$ has a P/O ratio of 2. The P/O ratio depends on the number of hydrogen atoms transported outward by an electrochemical gradient and the number of protons returned inward through the membrane by enzymes such as ATP synthase.

无效循环：也称底物循环，是由一对酶催化的循环反应，该循环通过 ATP 的水解释放热能。例如：葡萄糖+ATP = 葡萄糖-6-磷酸+ADP 与葡萄糖-6-磷酸+H_2O = 葡萄糖+Pi 组成的循环反应，其净反应实际上是 ATP+H_2O = ADP+Pi。葡萄糖代谢中的底物循环不是有效循环而是调节过程。例如，当突然需要能量时，ATP 会被更活泼的腺嘌呤 AMP 取代。

Futile cycle: Also known as the substrate cycle, which is a cycle reaction catalyzed by a pair of enzymes that results in the release of heat energy through the hydrolysis of ATP. For example, glucose+ATP = glucose-6-phosphate+ADP and glucose-6-phosphate+H_2O = glucose+Pi reaction composition of the cyclic reaction, the net reaction is actually ATP+H_2O=ADP+Pi. The substrate cycle of glucose metabolism is not an effective cycle but a regulatory process. For example, ATP is replaced by the more reactive adenine AMP when it suddenly needs energy.

酯交换是将一种羧酸酯转化为另一种羧酸酯的过程。在有机化学中，酯交换是将酯的有机基团 R″ 与醇的有机基团 R′ 交换的过程。这些反应通常需要添加酸或碱催化剂来催化，该反应也可以在酶，特别是脂肪酶的帮助下完成。图 2-1-5 所示为用于生物修复和生物燃料生产的综合藻类培养系统。

Transesterification is the conversion of one carboxylic ester to another. In organic chemistry, transesterification is the process of exchanging the organic group R″ of an ester with the organic group R′ of an alcohol. These reactions are usually catalyzed by the addition of acid or base catalysts, which can also be done with the help of enzymes, especially lipases. Fig. 2-1-5 shows integrated algae cultivation system for bioremediation and biofuel production.

第 2 章 工业生物技术

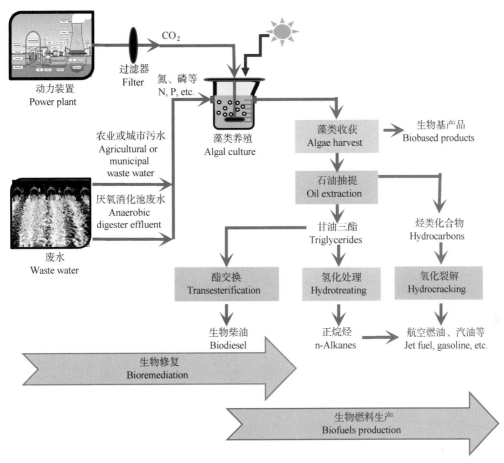

图 2-1-5 用于生物修复和生物燃料生产的综合藻类培养系统

Fig. 2-1-5 Integrated algae cultivation system for bioremediation and biofuel production

图 2-1-6 所示为原料生产工艺流程。图 2-1-6 (a) 所示为基于淀粉和糖类的第一代乙醇生产过程。由该图可见，淀粉可以水解成葡萄糖或果糖，并且可以进一步被发酵成乙醇。图 2-1-6 (b) 所示为木质纤维素乙醇的生产过程。木质纤维素生产乙醇的过程涉及生物质原料的预处理，水解产生单糖，以及发酵产生乙醇。木质纤维素乙醇被认为是第二代生物质能的主要成分。图 2-1-6 (c) 所示为生物柴油的生产过程。生物柴油的生产通常涉及脂肪酸的酯交换。图 2-1-6 (d) 所示为生物质气化的生产过程。生物质可通过气化过程产生甲醇、一氧化碳、氢气或其他气体。

Fig. 2-1-6 shows raw material production process. Fig. 2-1-6 (a) shows starch and sugar-based first-generation ethanol production process. In this platform, starch can be hydrolyzed into monosaccharides, which can be further fermented into ethanol. Fig. 2-1-6 (b) shows process lignocellulosic ethanol production process. Lignocellulosic ethanol production involves pretreatment of biomass material, hydrolysis for monosaccharide production, and fermentation to produce ethanol. Lignocellulosic ethanol is believed to be the major component of the second generation of bioenergy. Fig. 2-1-6 (c) shows biodiesel production process. Biodiesel production often involves the transesterification of fatty acids. Fig. 2-1-6 (d) shows biomass gasification production process. Biomass can be used to produce methanol, carbon monoxide, hydrogen or other gases through a process of gasification.

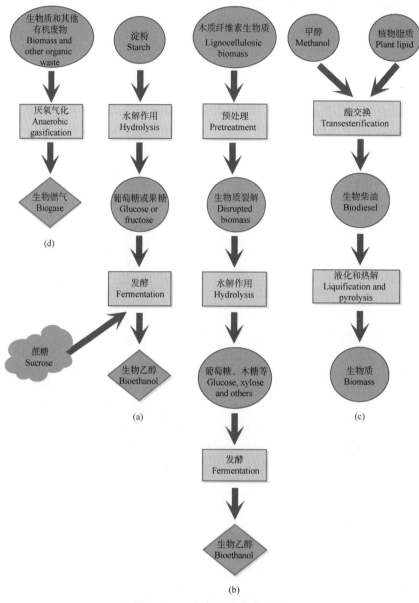

图 2-1-6 原料生产工艺流程

Fig. 2-1-6 Raw material production process

(a) 基于淀粉和糖类的第一代乙醇生产过程；(b) 木质纤维素乙醇生产过程；
(c) 生物柴油的生产过程；(d) 生物质气化生产过程

(a) Starch and sugar-based first-generation ethanol production process;
(b) Lignocellulosic ethanol production process; (c) Biodiesel production process;
(d) Biomass gasification production process

图 2-1-7 所示为利用各种原料生产不同生物燃料的加工平台。

Fig. 2-1-7 shows processing platforms for producing different biofuels with various feedstocks.

热化学转化：利用热和催化条件制造生物燃料和生物制品对于生物能源技术公司是一种经济的方式，可利用热能、压力和催化作用将常见非粮生物质转化为汽油、柴油、航空燃油以及其他产品。热化学转化过程可分为三种子类：热解、汽化和液化；生物质热化学转化技术主要是热解和汽化。

Thermochemical conversion: The manufacture of biofuels and biological products using thermal and catalytic conditions is an economical way for bioenergy technologies companies to develop the use of heat, pressure and catalysis to convert common non-food biomass into gasoline, diesel, jet fuel and other products. Thermochemical conversion processes include three subcategories: pyrolysis, gasification and liquefaction; biomass thermochemical conversion technology is mainly pyrolysis and gasification.

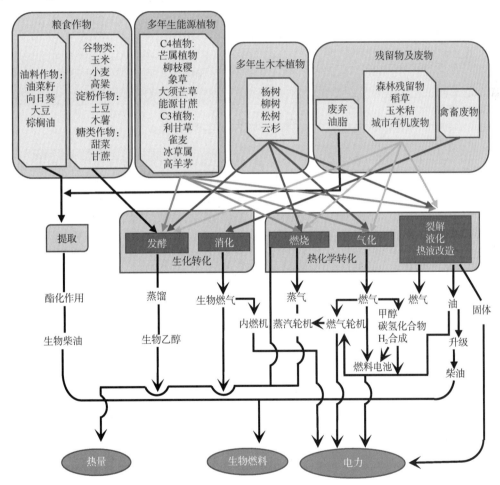

图 2-1-7 利用各种原料生产不同生物燃料的加工平台

生化转化：通过酶催化将木质纤维素原料水解为糖，随后通过发酵和蒸馏得到纤维素乙醇。生物化学方法具有加工温度低和产物选择性高的优点。生物质的生物化学转化涉及利用细菌、微生物和酶将生物质分解成气态或液态燃料，如生物燃气或生物乙醇。最普遍的生化技术是厌氧消化（或生物产甲烷）和发酵。

Biochemical conversion: the enzymatic hydrolysis of lignocellulosic materials to sugars, which are then fermented and distilled to produce cellulosic ethanol. Biochemical methods have the advantages of low processing temperature and high product selectivity. Biochemical conversion of biomass involves the use of bacteria, microorganisms and enzymes to break biomass down into gaseous or liquid fuels, such as biogas or bioethanol. The most common biochemical techniques are anaerobic digestion (or biomethanation) and fermentation.

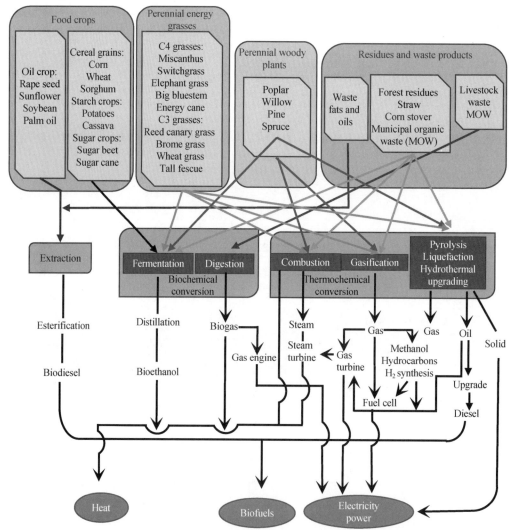

Fig. 2-1-7　Processing platforms for producing different biofuels with various feedstocks

厌氧消化：有机质在无氧条件下，由兼性菌和厌氧细菌将可生物降解的有机物分解为 CH_4、CO_2、H_2O 和 H_2S 的消化技术。厌氧消化被广泛应用于污水、畜禽粪便和城市有机废弃物处理，以及沼气工程技术中的其他方面。它可以实现循环经济发展、环境保护、减少温室气体排放和生产可再生能源等目标。食品和饮料的工业化生产以及家庭发酵大多采用厌氧消化。

Anaerobic digestion: It refers to the digestion technology in which facultative bacteria and anaerobic bacteria decompose biodegradable organic matter into CH_4, CO_2, H_2O and H_2S in the absence of oxygen. Anaerobic digestion is widely used in sewage, livestock and poultry excrement and urban organic waste treatment and other aspects of biogas engineering technology. It can achieve the development of circular economy, environmental protection, reduce greenhouse gas emissions and production of renewable energy and other goals. Most fermentations used industrially to produce food and beverages and in domestic fermentation use anaerobic digestion.

1.3 蛋白质的利用及代谢途径
1.3 Utilization and Metabolic Pathways of Proteins

生物精炼是指利用工业化生物技术加工处理生物质原料，将生物质转化为能源和其他有益副产品（如化学品）的综合技术和生产单位。图 2-1-8 所示为富含蛋白质的生物废弃物可作为未来生物精炼的资源。

Biorefining is a comprehensive technology and production unit that uses industrial biotechnology to process biomass feedstock and convert biomass into energy and other beneficial by-products (such as chemicals). Fig. 2-1-8 shows protein-rich biomass waste as a resource for future biorefinries.

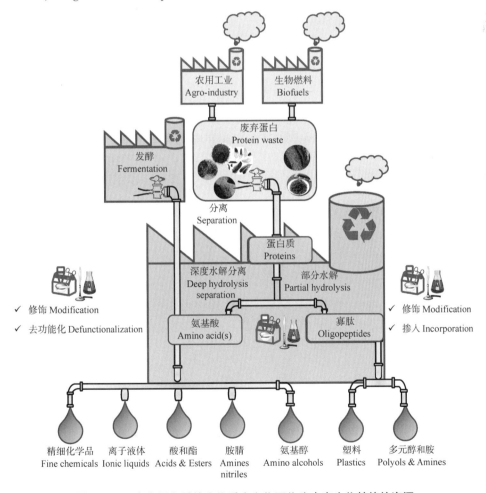

图 2-1-8 富含蛋白质的生物质废弃物可作为未来生物精炼的资源
Fig. 2-1-8 Protein-rich biomass waste as a resource for future biorefineries

图 2-1-9 所示为以氮为中心的大肠杆菌代谢工程策略。直接脱氨基的氨基酸未在该图中展示，其他氨基酸通过重编程的转氨和脱氨循环脱氨（A-D）。过表达的酶用红色显示。如图 2-1-9（a）所示，将天冬氨酸（Asp）和丙氨酸（Ala）中的氨基转移至 2-酮戊二酸（2-KG），通过一系列反应（未展示）得到丙酮酸（Pyr）和谷氨酸（Glu）。如图 2-1-9（b）所示，IlvE 将氨基从 Glu 转移至 2-酮甲基戊酸（KMV）或 2-酮异己酸（KIC），分别得到异亮氨酸（Ile）和亮氨酸（Leu）。如图 2-1-9（c）所示，IlvE 还可以将氨基从 Glu 转移到 2-酮异戊酸（KIV）并产生缬氨酸（Val）。然后 AvtA 将氨基从 Val 转移至 Pyr 以产生 L-Ala，然后将其转化为 D-Ala 并脱氨基得到 Pyr。如图 2-1-9（d）所示，SerC 将氨基从 Glu 转移至 3-磷酸羟基丙酮酸（3-PHP），得到 3-磷酸丝氨酸（3-P-Ser），然后通过 SerB 将其转化为丝氨酸（Ser）。Ser 被 SdaB 脱氨得到 Pyr。Pyr 可被回收生产 3-PHP 或用于燃料合成。如图 2-1-9（e）所示，工程酮酸途径生产乙醇（EtOH），异丁醇（iBOH）、2-甲基-1-丁醇（2MB）和 3-甲基-1-丁醇（3MB）。

Fig. 2-1-9 shows nitrogen-centric metabolic engineering strategy in *E. coli*. Amino acids that are directly deaminated are not shown in this figure. Others are deaminated through the reprogrammed transamination and deamination cycles (A-D). Overexpressed enzymes are shown in red. Fig. 2-1-9 (a) shows the amino groups in asparate (Asp) and alanine (Ala) are transferred to 2-ketoglutarate (2-KG) to yield pyruvate (Pyr) and glutamic acid (Glu) through a series of reactions (not shown). Fig. 2-1-9 (b) shows IlvE transfers the amino groups from Glu to 2-ketomethylvalerate (KMV) or 2-ketoisocaproate (KIC) to yield isoleucine (Ile) and leucine (Leu), respectively. Fig. 2-1-9 (c) shows IlvE also can transfer the amino groups from Glu to 2-ketoisovalerate (KIV) and generate valine (Val). AvtA then transfers the amino group from Val to Pyr to generate L-Ala, which is then converted to D-Ala and deaminated to yield Pyr. Fig. 2-1-9 (d) shows SerC transfers the amino groups from Glu to 3-phosphohydroxypyruvate (3-PHP) to yield 3-phosphoserine (3-P-Ser), which is then converted to serine (Ser) by SerB. Ser is deaminated by SdaB to generate Pyr. Pyr can be recycled to 3-PHP or used for fuel synthesis. Fig. 2-1-9 (e) shows engineered keto acid pathways that produce ethanol (EtOH), isobutanol (iBOH), 2-methyl-1-butanol (2MB) and 3-methyl-1-butanol (3MB).

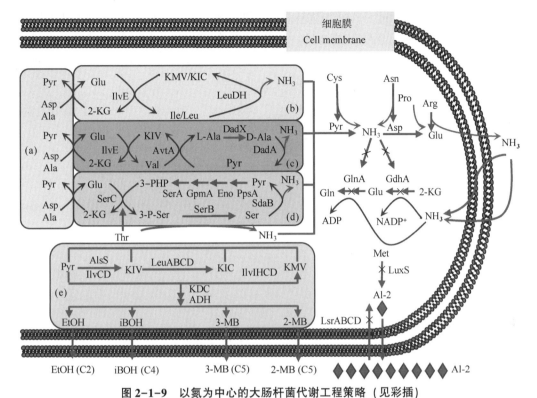

图 2-1-9 以氮为中心的大肠杆菌代谢工程策略（见彩插）
Fig. 2-1-9 Nitrogen-centric metabolic engineering strategy in *E. coli*（see the color figure）
（a）天冬氨酸和丙氨酸中的氨基转移到丙酮酸和谷氨酸；（b）谷氨酸氨基转移得到异亮氨酸和亮氨酸；
（c）谷氨酸氨基转移到丙酮酸；（d）谷氨酸中氨基转移到丙酮酸；（e）工程酮酸途径
(a) Amino groups in Asp and Ala are transferred to Pyr and Glu; (b) Amino groups in Glu are transferred to Ile and Leu;
(c) Amino groups are transferred to Pyr; (d) Amino groups in Glu are transferred to Pyr;
(e) Engineered keto acid pathways

图 2-1-10 所示为通过对寡肽和氨基酸的修饰和脱官能团，将富含蛋白质的生物质转化为一系列的大宗和精细化学品的过程。

Fig. 2-1-10 shows valorization of protein-rich biomass to a range of bulk and fine chemicals by modification and defunctionalization of oligopeptides and amino acids.

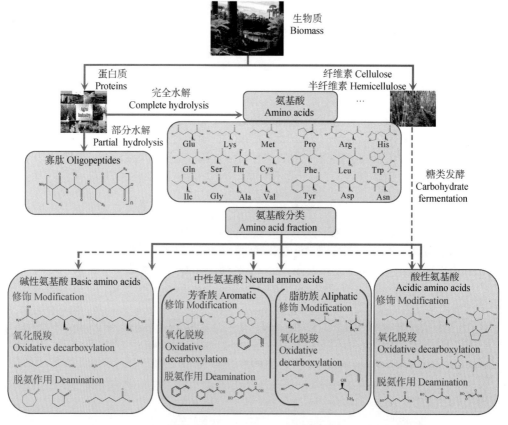

图 2-1-10 通过对寡肽和氨基酸的修饰和脱官能团，将富含蛋白质的生物质转化为一系列的大宗和精细化学品

Fig. 2-1-10 Valorization of protein-rich biomass to a range of bulk and fine chemicals by modification and defunctionalization of oligopeptides and amino acids

1.4 简单糖的利用及代谢途径
1.4 Utilization and Metabolic Pathways of Simple Sugar

图 2-1-11 所示为大肠杆菌将脂肪酸转化为燃料（红色）和化学品（绿色）的工程化途径。该图同时显示了脂肪酸经 β-氧化途径（橙色）和葡萄糖经 embden-meyerhof-parnas 途径（蓝色）的分解代谢。相关反应由编码酶的基因名称表示（未在括号中注明的基因均源自大肠杆菌，*ca* 表示丙酮丁醇梭菌，*cb* 表示拜氏梭菌）：*aceA* 表示异柠檬酸裂解酶；*aceB* 表示苹果酸合成酶 A；*adc* 表示乙酰乙酸脱羧酶（*ca*）；*ackA* 表示醋酸激酶；*adh* 表示仲醇脱氢酶（*cb*）；*adhE* 表示乙醛/乙醇脱氢酶；*adhE2* 表示仲醇脱氢酶（*ca*）；*atoA* 和 *atoD* 表示乙酰辅酶 A：乙酰辅酶 A 转移酶；*atoB* 表示乙酰辅酶 A 乙酰转移酶；*bcd* 表示丁酰辅酶 A 脱氢酶（*ca*）；*crt* 表示巴豆酶（*ca*）；*etfAB* 表示电子转移蛋白富集蛋白（*ca*）；*fadA* 表示 3-酮脂酰辅酶 A 硫解酶；*fadB* 表示烯酰辅酶 A 水合酶/3-羟基酰基辅酶 A 脱氢酶；*fadD* 表示酰基辅酶 A 合成酶；*fadE* 表示酰基辅酶 A 脱氢酶；*hbd* 表示 β-羟基丁酰基-CoA 脱氢酶（*ca*）；*icd* 表示异柠檬酸脱氢酶；*pta* 表示磷酸乙酰转移酶；*sdhABCD* 表示琥珀酸脱氢酶；*scpA* 表示甲基丙二酰辅酶 A 变位酶；*scpB* 表示甲基丙二酰辅酶 A 脱羧酶；*scpC* 表示丙酰辅酶 A：琥珀酸辅酶 A 转移酶；*sucA* 表示 2-氧代戊二酸脱氢酶；*sucB* 表示二氢硫辛酸琥珀酰基转移酶；*sucCD* 表示琥珀酰辅酶 A 合成酶。

缩写：2 [H] = NADH = FADH$_2$ = QH$_2$ = H$_2$；*P/O*，氧化磷酸化中每消耗一个氧气产生的 ATP 量。

Fig. 2-1-11 shows pathways engineered in *E. coli* for the conversion of fatty acids to fuels (red) and chemicals (green). It also shows the catabolism of fatty acids via the β-oxidation pathway (orange) and of glucose through the embden-meyerhof-parnas pathway (blue). Relevant reactions are represented by the names of the genes coding for the enzymes (*E. coli* genes unless otherwise specified in parentheses as follows: *C. acetobutylicum*, *ca*; *C. beijerinckii*, *cb*): *aceA*, isocitrate lyase; *aceB*, malate synthase A; *adc*, acetoacetate decarboxylase (*ca*); *ackA*, acetate kinase; *adh*, secondary alcohol dehydrogenase (*cb*); *adhE*, acetaldehyde/alcohol dehydrogenase; *adhE2*, secondary alcohol dehydrogenase (*ca*); *atoA* and *atoD*, acetyl-CoA: acetoacetyl-CoA transferase; *atoB*, acetyl-CoA acetyltransferase; *bcd*, butyryl-CoA dehydrogenase (*ca*); *crt*, crotonase (*ca*); *etfAB*, electron transfer flavoprotein (*ca*); *fadA*, 3-ketoacyl-CoA thiolase; *fadB*, enoyl-CoA hydratase/3-hydroxyacyl-CoA dehydrogenase; *fadD*, acyl-CoA syn-thetase; *fadE*, acyl-CoA dehydrogenase; *hbd*, β-hydroxybutyryl-CoA dehydrogenase (*ca*); *icd*, isocitrate dehydrogenase; *pta*, phosphate acetyl-transferase; *sdhABCD*, succinate dehydrogenase; *scpA*, methylmalonyl-CoA mutase; *scpB*, methylmalonyl-CoA decarboxylase; *scpC*, propionyl-CoA: succinate CoA transferase; *sucA*, 2-oxoglutarate dehydrogenase; *sucB*, dihydrolipoyltranssuccinylase; and *sucCD*, succinyl-CoA synthetase.

Abbreviations: 2 [H] = NADH = FADH$_2$ = QH$_2$ = H$_2$; *P/O*, amount of ATP produced per oxygen consumed in the oxidative phosphorylation.

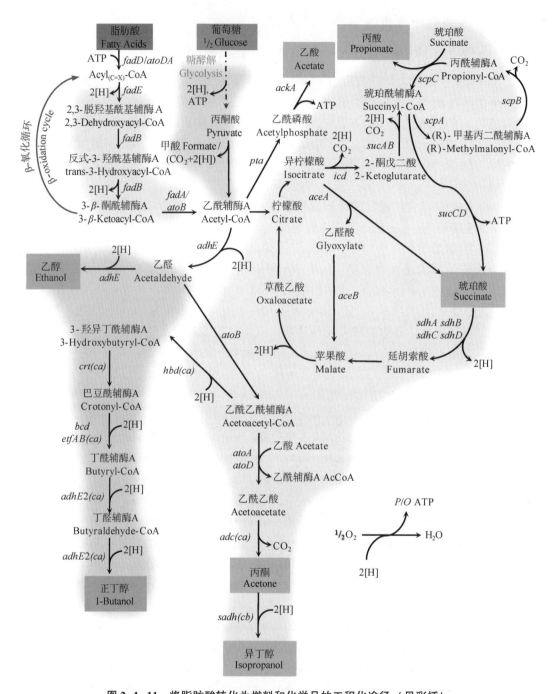

图 2-1-11 将脂肪酸转化为燃料和化学品的工程化途径（见彩插）
Fig. 2-1-11 Pathways engineered in *E. coli* for the conversion of fatty acids to fuels and chemicals (see the color figure)

1.5 木糖的利用及代谢途径
1.5 Utilization and Metabolism Pathways of Xylose

图 2-1-12 所示木糖代谢途径：蓝实线为 Weimberg 途径；蓝虚线为 Dahms 途径；红实线为异构酶途径；绿实线为氧化还原酶途径。

Fig. 2-1-12 shows metabolic pathway of xylose：solid blue is Weimberg pathway；dashed blue is dahms pathway；solid red is isomerase pathway；solid green is oxidoreductase pathway.

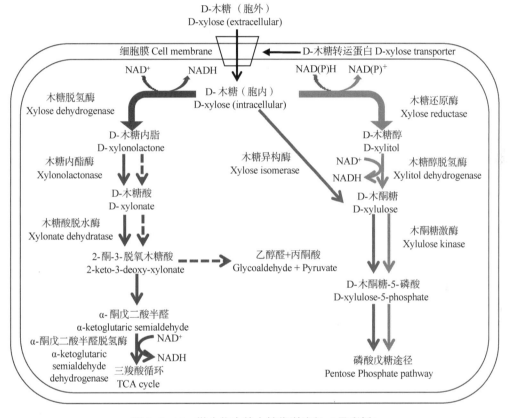

图 2-1-12　微生物中的木糖代谢途径（见彩插）

Fig. 2-1-12　Xylose metabolic pathways in microorganisms（see the color figure）

图 2-1-13 所示为菌株中工程木糖同化途径和中心碳代谢蓝细菌集胞藻 6803 菌株中工程木糖同化途径和中心碳代谢过程。碳源（木糖和/或 CO_2）进入带有 xylA、xylB 基因的工程化集胞藻菌株的中心碳代谢。通过 Calvin-Benson-Bassham（CBB）循环进行的光合 CO_2 固定产生代谢中间体碳化合物。当木糖通过内源性木糖转运蛋白进入细胞时，首先被木糖异构酶（XI）异构化为木酮糖，然后被木酮糖激酶（XK）磷酸化，生成戊糖磷酸化途径（PPP）的中间体木糖-5-磷酸（X-5-P）。对于工程化乙烯生产，乙烯生成酶（EFE）催化 TCA 循环中间体 α-酮戊二酸（AKG）向乙烯的转化。在糖原合成受阻的 ΔglgC 菌株中会发生溢流代谢，产生 AKG 和丙酮酸。

Fig. 2-1-13 shows overview of the engineered xylose assimilation pathway and central carbon metabolism in *Synechocystis* 6803 strain. Carbon sources（xylose and/or CO_2）enter central carbon metabolism in the engineered *Synechocystis* strains bearing *xylAB* genes. Photosynthetic CO_2 fixation via the Calvin-Benson-Bassham (CBB) cycle produces metabolic intermediate carbon compounds. When xylose enters cells via an endogenous xylose transporter, it is first isomerized into xylulose by xylose isomerase (XI) and then phosphorylated by xylulokinase (XK) to produce xylulose-5-phosphate (X-5-P), an intermediate in the pentose phosphate pathway (PPP). For engineered ethylene production, ethylene forming enzyme (EFE) catalyzes the conversion of the TCA cycle intermediate α-ketoglutarate (AKG) into ethylene. AKG and pyruvate production by overflow metabolism occurs in Δ*glgC* strains with blocked glycogen synthesis.

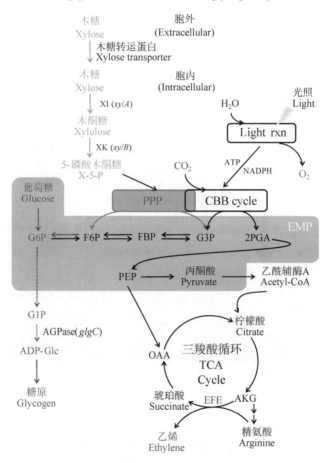

图 2-1-13 蓝细菌集胞藻菌株中工程木糖同化途径和中心碳代谢

Fig. 2-1-13 Engineered xylose assimilation pathway and central carbon metabolism in *Synechocystis*

注释：
G1P，葡萄糖-1-磷酸；ADP-Glc，腺苷二磷酸-葡萄糖；G3P，3-磷酸甘油醛；EMP，embden-meyerhof-parnas 途径；rxn，反应；2PGA，2-磷酸甘油酸；PEP，磷酸烯醇丙酮酸；OAA，草酰乙酸；AKG，α-酮戊二酸；F6P，果糖-6-磷酸；FBP，果糖-1, 6-二磷酸；G6P，6-磷酸葡萄糖；AGPase，ADP 葡萄糖焦磷酸化酶。

Note：
G1P, glucose-1-phosphate; ADP-Glc, ADP-glucose; G3P, glyceraldehyde-3-phosphate; EMP, embden-meyerhof-parnas pathway; rxn, reaction; 2PGA, 2-phosphoglycerate; PEP, phosphoenolpyruvate; OAA, oxaloacetic acid; AKG, α-ketoglutarate; F6P, fructose-6-phosphate; FBP, fructose-1, 6-bis-phosphate; G6P, glucose-6-phosphate; and AGPase, ADP glucose pyrophosphorylase.

1.6 木质纤维素的利用及代谢途径
1.6 Utilization and Metabolic Pathways of Lignocellulose

图 2-1-14（a）所示为工程酵母通过整合三种重要纤维素生物质组分（纤维二糖，木糖和乙酸）的共同利用途径来生产乙醇；图 2-1-14（b）所示为纤维素的降解途径。

Fig. 2-1-14（a）shows engineered yeast produces ethanol by integrating the pathways for co-utilization of three important cellulosic biomass components（cellobiose, xylose, and acetic acid）; Fig. 2-1-14（b）shows degradation pathway of cellulose.

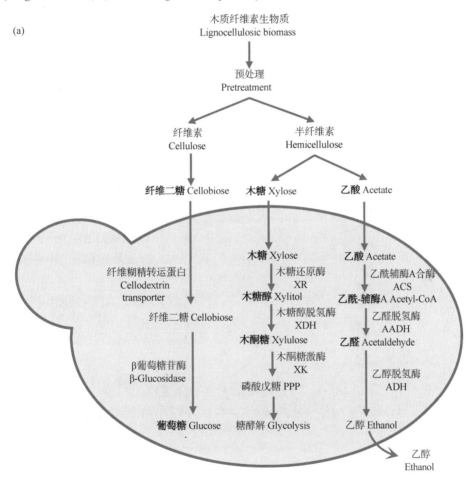

图 2-1-14　木质纤维素的利用及代谢途径
Fig. 2-1-14　Utilization and metabolic pathways of lignocellulose
（a）工程酵母通过整合三种重要纤维素生物质组分
（a）Engineered yeast produces ethanol by integrating the pathways of three important cellulosic biomass components

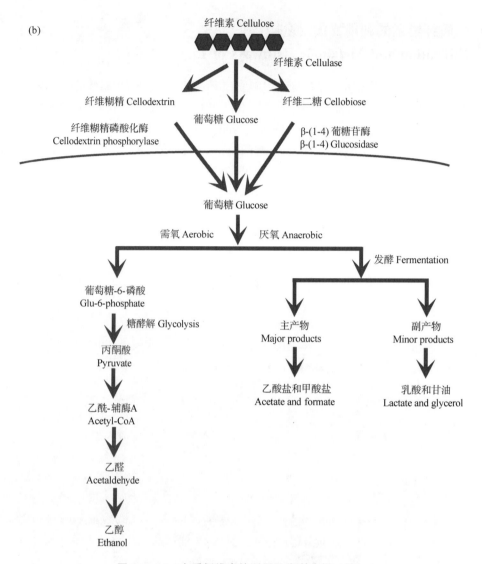

图 2-1-14　木质纤维素的利用及代谢途径（续）

Fig. 2-1-14　Utilization and metabolic pathways of lignocellulose (continued)

(b) 纤维素的降解途径

(b) Degradation pathway of cellulose

1.7 CO_2的利用及代谢途径
1.7 Utilization and Metabolic Pathways of CO_2

图 2-1-15（a）所示为碳循环，图 2-1-15（b）所示为真核生物（①）和原核生物（②）中一些次生和初级（碳水化合物，淀粉，酒精等）代谢物的主要生物合成途径。

Fig. 2-1-15（a）shows carbon cycle；Fig. 2-1-15（b）shows main pathways for the biosynthesis of some secondary and well-primary (carbohydrate, starch, alcohol, etc.) metabolites in eukaryote（①）and prokaryote（②）.

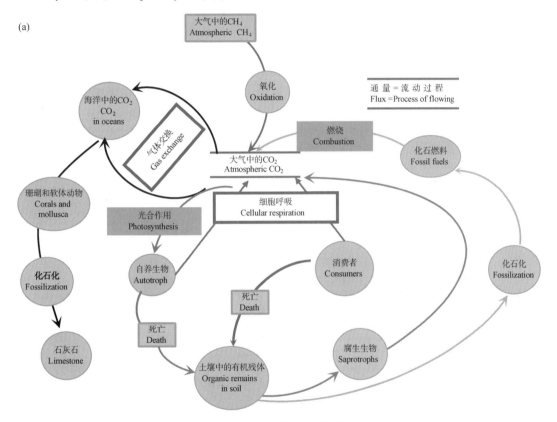

图 2-1-15 CO_2 的利用及代谢途径

Fig. 2-1-15 Utilization and metabolic pathway of CO_2

（a）碳循环

（a）Carbon cycle

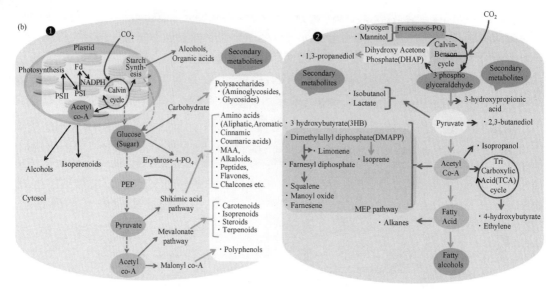

图 2-1-15　CO_2 的利用及代谢途径（续）

Fig. 2-1-15　Utilization and metabolic pathway of CO_2（continued）

（b）真核生物和原核生物中一些次生和初级代谢物的主要生物合成途径

（b）Main pathways for the biosynthesis of some secondary and well-primary metabolites in eukaryote and prokaryote

注释：

Note：

Plastid，质体；Photosynthesis，光合作用；CO_2，二氧化碳；Acetyl-CoA，乙酰辅酶 A；Alcohol，醇；Isoprenoids，类异戊二烯；Cytosol，胞质；Calvin cycle，卡尔文循环；Starch synthesis，淀粉合成；Glucose，葡萄糖；PEP，磷酸烯醇式丙酮酸；Pyruvate，丙酮酸；Organic acids，有机酸；Carbohydrate，碳水化合物；Erythrose-4-PO_4，赤藓糖-4-磷酸；Shikimic acid pathway，莽草酸途径；Mevalonate pathway，甲羟戊酸途径；Malonyl CoA，丙二酰辅酶 A；Secondary metabolites，次级代谢产物；Polysaccharides，多糖；Aminoglycosides，氨基糖苷类；Glycosides，糖苷；Amino acid，氨基酸；Aliphatic，脂肪族；Aromatic，芳香族；Cinnamic，肉桂类；Coumaric acids，香豆酸；MAA，甲基丙烯酸；Alkaloids，生物碱；Peptides，肽类；Flavones，黄酮类；Chalcones，查耳酮；Carotenoids，类胡萝卜素；Isoprenoids，类异戊二烯；Steroids，类固醇；Terpenoids，萜类化合物；Polyphenols，多酚类物质；Glycogen，糖原；Mannitol，甘露醇；1,3-propanediol，1,3-丙二醇；Fructose-6-PO_4，6-磷酸果糖；DHAP，磷酸二羟基丙酮；3-phospho glyceraldehyde，3-磷酸甘油醛；Isobutanol，异丁醇；Lactate，乳酸；3HB，3-羟基丁酸；DMAPP，二甲烯丙基二磷酸；Limonene，柠檬烯；Isoprene，异戊二烯；Farnesyl diphosphate，法呢基二磷酸；Squalene，角鲨烯；Manoyl oxide，泪柏醚；Farnesene，法呢烯；Alkanes，烷烃；3-hydroxypropionic acid，3-羟基丙酸；2,3-butanediol，2,3-丁二醇；Isopropanol，异丙醇；TCA cycle，三羧酸循环；Fatty acid，脂肪酸；Fatty alcohols，脂肪醇；4-hydroxybutyrate，4-羟基丁酸酯；Ethylene，乙烯。

1.8　原料利用能力的强化
1.8　Enhancement of Raw Material Utilization

Calvin-Benson-Bassham 循环是二氧化碳固定最重要的途径。图 2-1-16（a）所示为 CBB 途径中的酶和中间体的代谢图。图 2-1-16（b）所示为 RuBisCO 大亚基序列的系统发生树。不同的 RuBisCO 类别以颜色区分：类型 I（黑色），类型 II（蓝色）和古菌的类型 III（红色）。

The Calvin-Benson-Bassham cycle is the most significant pathway for CO_2 fixation. Fig. 2-1-16

(a) Metabolic diagram depicting the enzymes and intermediates in the CBB pathway. Fig. 2-1-16 (b) phylogenetic tree of RuBisCO large subunit sequences. Different RuBisCO classes are colored: Form I (black), Form II (blue), and the archeal Form III (red).

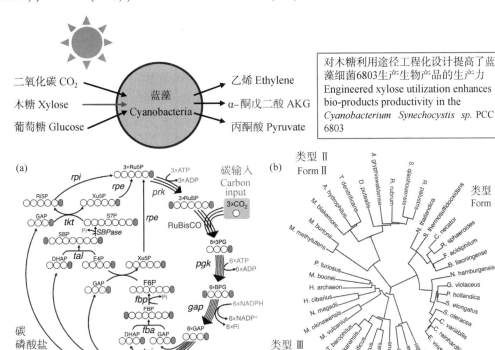

图 2-1-16 木糖利用途径工程化设计（见彩插）
Fig. 2-1-16 Engineered xylose utilization (see the color figure)
(a) CBB 途径中的酶和中间体的代谢；(b) RuBisCO 大亚基序列的系统发生树
(a) Metabolic diagram of the enzymes and intermediates in the CBB pathway;
(b) Phylogenetic tree of RuBisCO large subunit sequences

注释：
RuBP, 1, 5-二磷酸核酮糖；Ru5P, 5-磷酸核酮糖；3PG, 3-磷酸甘油酸；BPG, 1, 3-二磷酸甘油酸；GAP, 甘油醛 3-磷酸；DHAP, 磷酸二羟丙酮；FBP, 1, 6-二磷酸果糖；F6P, 6-磷酸果糖；Xu5P, 5-磷酸木酮糖；E4P, 4-磷酸赤藓糖；SBP, 1, 7-二磷酸景天庚酮糖；S7P, 7-磷酸景天庚酮糖；Ri5P, 5-磷酸核糖；*rpi*：5-磷酸核糖异构酶；*rpe*：磷酸核酮糖差向异构酶；*pgk*：磷酸甘油酸激酶；*gap*：甘油醛-3-磷酸脱氢酶；*tpi*：磷酸丙糖异构酶；*fba*：1, 6-二磷酸果糖醛缩酶；*fbp*：果糖 1, 6-二磷酸；*tal*：转醛醇酶；*SBPase*：景天庚酮糖二磷酸酶；*tkt*：转酮醇酶。

Note：
RuBP, ribulose-1, 5-bisphosphate; Ru5P, ribulose-5-phosphate; 3PG, 3-phosphoglycerate; BPG, 1, 3-bisphosphoglycerate; GAP, glyceraldehyde 3-phosphate; DHAP, dihydroxyacetone phosphate; FBP, fructose 1, 6-bisphosphate; F6P, fructose 6-phosphate; Xu5P, xylose 5-phosphate; E4P, erythritose 4-phosphate; SBP, sedoheptulose-1, 7-bisphosphate; S7P, sedoheptulose-7-phosphate; Ri5P, ribose-5-phosphate; *rpi*: ribose-5-phosphate isomerase; *rpe*: ribulose-phosphate 3-epimerase; *pgk*: phosphoglycerate kinase; *gap*: glyceraldehyde-3-phosphate dehydrogenase; *tpi*: triosephosphate isomerase; *fba*: fructose-1, 6-bisphosphate aldolase; *fbp*: fructose 1, 6-bisphosphatase; *tal*: transaldolase; *SBPase*: sedoheptulose-bisphosphatase; *tkt*: transketolase.

Ⅱ 产物的合成
Ⅱ Product Synthesis

2.1 生物燃料合成
2.1 Biofuel Synthesis

图 2-2-1 所示为工程化大肠杆菌中生产高级醇的合成网络。红色箭头代表 2-酮酸脱羧和还原途径。蓝色的酶名称代表氨基酸生物合成途径。

Fig. 2-2-1 shows the synthetic networks for the higher alcohols production in engineered *E. coli*. Red arrows represent the 2-keto acid decarboxylation and reduction pathway. Blue enzyme names represent amino acid biosynthesis pathways.

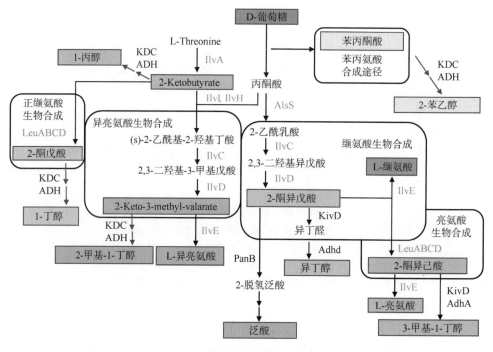

图 2-2-1 工程化大肠杆菌中生产高级醇的合成网络（见彩插）

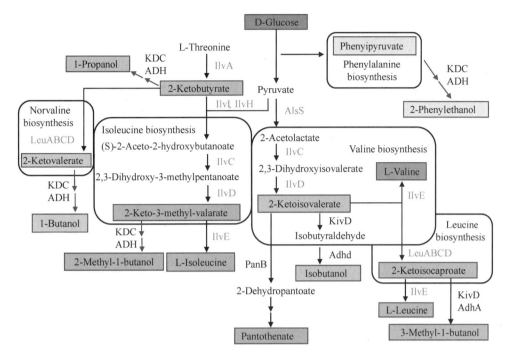

Fig. 2-2-1 The synthetic networks for the higher alcohols production in engineered *E. coli*. (see the color figure)

注释：

Note：

IlvA，苏氨酸脱氨酶；IlvC，酮酸还原异构酶；IlvD，二羟基酸脱水酶；IlvH，乙酰乳酸合成酶；IlvE，链氨基酸转移酶；AlsS，乙酰乳酸合成酶；KivD，α-酮异戊酸脱羧酶；Adh，乙醇脱氢酶。

高级醇：具有超过2个碳原子的醇（乙醇具有两个碳原子），因而具有更高的分子量和更高的沸点。

Higher alcohols: alcohols that have more than 2 carbons (ethanol has two carbons) and thus have higher molecular weight and higher boiling point.

异丁醇：一种分子式为 $(CH_3)_2CHCH_2OH$ 的有机化合物。这种无色易燃液体主要用作化学反应的溶剂。其异构体如正丁醇、2-丁醇和叔丁醇都具有重要的工业价值。

Isobutanol: an organic compound with the formula $(CH_3)_2CHCH_2OH$. This colorless, flammable liquid is mainly used as a solvent for chemical reactions. Its isomers, include n-butanol, 2-butanol, and tert-butanol, all of which are important industrially.

2.2 芳香族化合物合成
2.2 Aromatic Compound Synthesis

图 2-2-2 所示为莽草酸途径（以绿色显示）和选定的代谢物。分支酸是植物和微生物中重要的生化中间体，是色氨酸途径（蓝色）和苯丙氨酸/酪氨酸途径（红色）的共同前体分子。

Fig. 2-2-2 shows the shikimic acid pathway (shown in green) and selected metabolites. Chorismate is an important biochemical intermediate in plants and microorganisms. It is the common precursor molecule for the tryptophan pathway (blue) and the phenylalanine/tyrosine pathways (red).

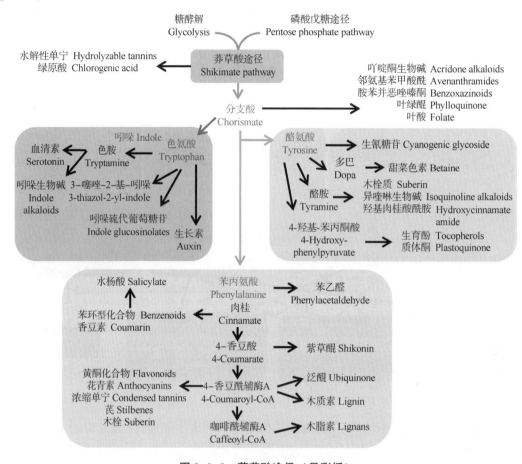

图 2-2-2 莽草酸途径（见彩插）
Fig. 2-2-2 The shikimic acid pathway (see the color figure)

2.3 天然产物合成
2.3 Natural Product Synthesis

图 2-2-3 所示为植物细胞溶质和质体中的已知和新型萜烯的生物合成途径。萜烯前体通过甲羟戊酸（MVA）途径（细胞溶质）和甲基赤藓糖醇磷酸（MEP）途径（质体）合成。

Fig. 2-2-3 shows known and novel terpene biosynthesis pathways in plant cytosol and plastids.

Terpene precursors are synthesized by themevalonic acid (MVA) pathway (cytosolic) and the methylerythritol phosphate (MEP) pathway (plastidic).

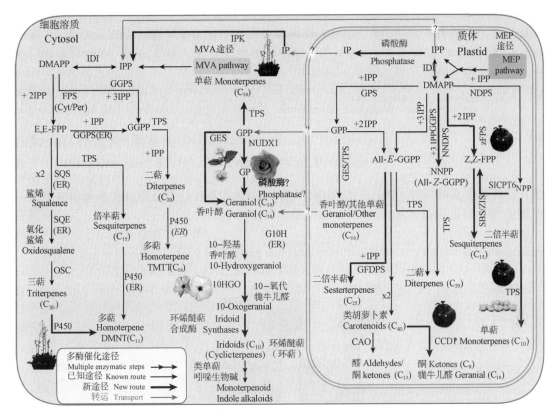

图 2-2-3　植物细胞溶质和质体中的已知和新型萜烯的生物合成途径
Fig. 2-2-3　Known and novel terpene biosynthesis pathways in plant cytosol and plastids

注释：
ER，内质网；Per，过氧化物酶体；CAO，类胡萝卜素加氧酶；CCD1，类胡萝卜素裂解双加氧酶 1；DMAPP，二甲烯丙基焦磷酸；DMNT，4，8-二甲基壬-1，3，7-三烯；E，E/Z，Z-FPP，所有 E/Z 法尼基焦磷酸酯；FPS，E，E-FPP 合成酶；zFPS，Z，Z-FPP 合成酶；G10H，香叶醇 10-羟化酶；10HGO，10-羟基香叶醇脱氢酶；GES，香叶醇合成酶；GFDPS，香叶基法尼基焦磷酸合酶；GP，单磷酸香叶基酯；GPP，香叶基二磷酸；GPS，GPP 合成酶；All-E-GGPP，(全部 E) 牻牛儿基焦磷酸；GGPS，GGPP 合成酶；IDI，异戊烯基二磷酸异构酶；IP，异戊烯磷酸酯；IPK，激酶；IPP，异戊烯焦磷酸；NDPS，二磷酸二氢萘合成酶；NNPP，二苯乙烯基二磷酸酯；NNDPS，二苯乙烯基二磷酸酯合酶；NPP，二十二烷基磷酸酯；NUDX1，Nudix 水解酶 1；OSC，氧化角鲨烯环化酶；P450，细胞色素 P450 单加氧酶；SBS，檀香烯和佛手柑合酶；SlCPT6，番茄茄属植物的 Z-异戊烯基转移酶；SQE，角鲨烯环氧酶；SQS，角鲨烯合成酶；TMTT，4，8，12-三甲基三嗪-1，3，7，11-四烯；TPS，萜烯合成酶；ZIS，d-姜烯合成酶。

Note：
ER, endoplasmic reticulum；Per, peroxisome；CAO, carotenoid oxygenase；CCD1, carotenoid cleavage dioxygenase 1；DMAPP, dimethylallyl diphosphate；DMNT, 4, 8-dimethylnona-1, 3, 7-triene；E, E/Z, Z-FPP, all E/Z farnesyl diphosphate；FPS, E, E-FPP synthase；zFPS, Z, Z-FPP synthase；G10H, geraniol 10-hydroxylase；10HGO, 10-hydroxygeraniol dehydrogenase；GES, geraniol synthase；GFDPS, geranylfarnesyl diphosphate synthase；GP, geranyl monophosphate；GPP, geranyl diphosphate；GPS, GPP synthase；(All-E)-GGPP, (all E) geranylgeranyl diphosphate；GGPS, GGPP synthase；IDI, isopentenyl diphosphate isomerase；IP, isopentenyl phosphate；IPK, IP kinase；IPP, isopentenyl diphosphate；NDPS, neryl diphosphate synthase；NNPP, nerylneryl diphosphate；NNDPS, nerylneryl diphosphate synthase；NPP, neryl diphosphate；NUDX1, Nudix hydrolase 1；OSC, oxidosqualene cyclase；P450, cytochrome P450 monooxygenase；SBS, santalene and bergamotene synthase；SlCPT6, Z-prenyltransferase 6 of *Solanum lycopersicum*；SQE, squalene epoxidase；SQS, squalene synthase；TMTT, 4, 8, 12-trimethyltrideca-1, 3, 7, 11-tetraene；TPS, terpene synthase；ZIS, d-zingiberene synthase.

甲羟戊酸途径：也称为类异戊二烯途径或 HMG-CoA 还原酶途径，是存在于真核生物、古细菌和一些细菌中的重要代谢途径。该途径产生两个称为异戊烯焦磷酸 IPP 和二甲基烯丙基焦磷酸 DMAPP 的五碳结构单元，用于合成类异戊二烯，是类固醇、类萜等生物分子的合成前体。

MVA pathway (mevalonate pathway): also known as the isoprenoid pathway or HMG-CoA reductase pathway, which is an essential metabolic pathway present in eukaryotes, archaea, and some bacteria. The pathway produces two five-carbon building blocks called isopentenyl pyrophosphate IPP and dimethylallyl pyrophosphate DMAPP, which are used to make isoprenoids.

甲基赤藓糖醇-4-磷酸途径：异戊二烯前体异戊烯焦磷酸 IPP 和二甲基烯丙基焦磷酸 DMAPP 生物合成的替代合成途径。甲基赤藓糖醇-4-磷酸是通往异戊烯焦磷酸的第一个代谢产物。

MEP pathway (methylerythritol-4-phosphate pathway): an alternative metabolic pathway for the biosynthesis of the isoprenoid precursors isopentenyl pyrophosphate IPP and dimethylallylpyrophosphate DMAPP. MEP is the first committed metabolite on the route to IPP.

图 2-2-4 (a) 所示为通过 PKS 实现聚酮化合物链延伸的一般反应流程；图 2-2-4 (b) 所示为通过聚酮化合物合酶 PKS 途径生物合成二十二碳六烯酸 DHA。AT：酰基转移酶；DH：脱水酶；ACC：乙酰辅酶 A 羧化酶；ACP：酰基载体蛋白；MT：丙二酰基转移酶；KS：酮酰合成酶；KR：酮还原酶；D/I：脱水酶/异构酶；ER：烯酰还原酶；图 2-2-4 (c) 所示为酚类脂质的生物合成途径；图 2-2-4 (d) 所示为维司他丁的生物合成途径。

Fig. 2-2-4 (a) shows general reaction scheme of polyketide chain elongation by PKS; Fig. 2-2-4 (b) shows polyketide synthase PKS pathways for docosahexaenoic acid (DHA) biosynthesis. AT: acyltransferase; DH: dehydratase; ACC: Acetyl-CoA carboxylase; ACP: Acyl carrier protein; MT: Malonyl transferase; KS: Ketoacyl synthase; KR: Keto reductase; D/I: Dehydrase/isomerase; ER: Enoyl reductase; Fig. 2-2-4 (c) shows the biosynthetic pathway of phenolic lipids; Fig. 2-2-4 (d) shows the biosynthetic pathway of vicenistatin.

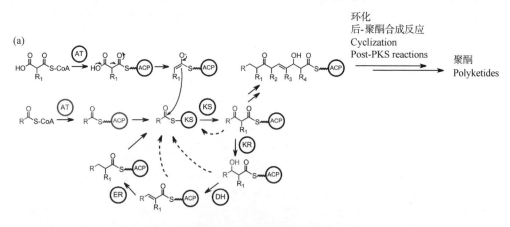

图 2-2-4 生物合成途径

Fig. 2-2-4 The biosynthetic pathway

(a) 通过 PKS 实现聚酮化合物链延伸的一般反应流程

(a) General reaction scheme of polyketide chain elongation by PKS

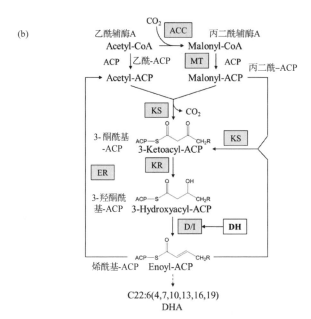

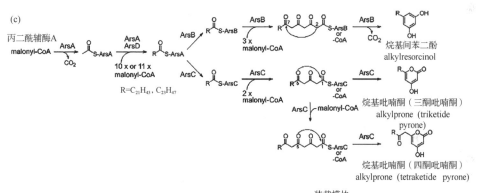

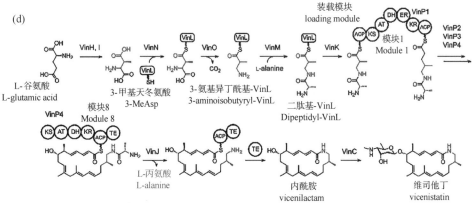

图 2-2-4 生物合成途径（续）

Fig. 2-2-4 The biosynthetic pathway (continued)

（b）通过聚酮化合物合酶 PKS 途径生物合成二十二碳六烯酸 DHA；
（c）酚类脂质的生物合成途径；（d）维司他丁的生物合成途径

(b) Polyketide synthase PKS pathways for docosahexaenoic acid (DHA) biosynthesis；
(c) Biosynthetic pathway of phenolic lipids；(d) Biosynthetic pathway of vicenistatin

聚酮化合物是一大类次级代谢物，含有交替的羰基和亚甲基（—CO—CH$_2$），或者衍生自含有这种交替基团的前体。许多聚酮化合物可药用或具有急性毒性。它们在细菌、真菌、植物和某些海洋动物中产生。生物合成涉及乙酰辅酶 A 或丙酰辅酶 A 与丙二酰辅酶 A 或甲基丙二酰辅酶 A 的逐步缩合。缩合反应伴随着延伸单元的脱羧并产生 β-酮官能团。第一次缩合产生乙酰基，即二酮化合物。随后的缩合产生三酮、四酮等化合物。

Polyketides are a large group of secondary metabolites which either contain alternating carbonyl groups and methylene groups (—CO—CH$_2^-$), or are derived from precursors which contain such alternating groups. Many polyketides are medicinal or exhibit acute toxicity. They are produced in bacteria, fungi, plants, and certain marine animals. The biosynthesis involves stepwise condensation of acetyl-CoA or propionyl-CoA with either malonyl-CoA or methylmalonyl-CoA. The condensation reaction is accompanied by the decarboxylation of the extender unit and yields a β-keto functional group. The first condensation yields an acetoacetyl group, a diketide. Subsequent condensations yield triketides, tetraketide, etc.

2.4 聚合物合成
2.4 Polymer Synthesis

图 2-2-5 所示为通过生化转化将木质素转化为聚羟基链烷酸酯和粘康酸。

Fig. 2-2-5 shows biochemical transformation of lignin to polyhydroxyalkanoates and muconic acid.

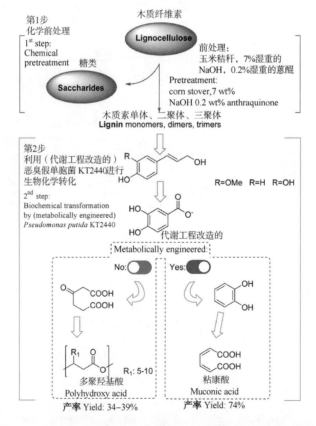

图 2-2-5 通过生化转化将木质素转化为聚羟基链烷酸酯和粘康酸

Fig. 2-2-5 Biochemical transformation of lignin to polyhydroxyalkanoates and muconic acid

聚羟基脂肪酸（PHAs）：细菌发酵糖或脂质产生的聚酯，可由多种微生物产生。它们既可以作为能量来源，也可以作为贮存碳源。

Polyhydroxyalkanoates (PHAs): polyesters produced by numerous microorganisms through bacterial fermentation of sugars or lipids. They serve as both a source of energy and as a carbon store.

Ⅲ 底盘的改造
Ⅲ Modification of Microbial Chassis

3.1 常见底盘
3.1 Common Chassis

常见底盘概况如表 2-3-1 所示。

Table 2-3-1 shows common chassis overview.

表 2-3-1 常见底盘概况
Table 2-3-1 Common chassis overview

常见底盘 Common chassis	特点 Features	菌落形态 Colony morphology
大肠杆菌 *Escherichia coli*	遗传背景清晰，载体受体系统完备，生长迅速，容易培养；缺乏蛋白质加工系统，目标基因表达水平高；表达系统成熟完善；可产生结构复杂和种类繁多的内毒素 Clear genetic background, complete vector receptor system, rapid growth and easy to cultivate; Lack of protein processing system, high target gene expression level; Mature and perfect expression system; Production of complex structures and a wide variety of endotoxins	
酵母菌 *Yeast*	具有真核生物的特征，遗传背景清晰，生长迅速，容易培养，外源基因表达系统完善，遗传稳定；内源性蛋白产物种类多且含量高；适用于外源 DNA 的扩增，克隆及真核生物基因的高效表达，基因文库的构建，真核生物基因表达调控的研究 It has the characteristics of eukaryotes, with clear genetic background, rapid growth, easy cultivation, perfect foreign gene expression system and genetic stability; Endogenous protein products are diverse and high in content. It is suitable for the amplification of exogenous DNA, cloning and high-efficiency expression of eukaryotic genes, the construction of gene libraries, and the study of eukaryotic gene expression regulation	

续表

常见底盘 Common chassis		特点 Features	菌落形态 Colony morphology
谷氨酸棒杆菌 *Corynebacterium glutamicum*		是一种对人类安全的菌株；可快速生长至高细胞密度；由于缺乏重组修复系统而在遗传上稳定；在生长和污染的情况下不会自动分解并维持代谢活动；代谢可塑性和强健的次级代谢特性 It is a safe strain for humans; it grows quickly to high cell density; It is genetically stable due to the lack of a recombination repair system; It does not automatically decompose and maintain metabolic activities in the case of growth and contamination; metabolic plasticity and strong secondary metabolic properties	
乳酸菌 *Lactobacillus*	乳酸球菌 *Lactococcus*	无强抗原性；分泌蛋白较少，减少了载体微生物的自身蛋白对外源分泌蛋白的干扰；不产生任何胞外蛋白酶，不会引起分泌蛋白发生胞外降解 No strong antigenicity; Less secreted protein, which reduces the interference of the carrier microorg-anism's own protein to exogenous secreted protein; It does not produce any extracellular proteases, and will not cause the secreted protein to undergo extracellular degradation	
	乳酸杆菌 *Lactobacillus*	能在呼吸系统、消化系统、泌尿系统定植，对维持微生态平衡具有重要作用；作为食品级细菌，易构建成食品级的基因克隆及表达系统，可以有效提高基因工程产品的安全性，并可在一定程度上简化表达产物的后期处理工艺；与肠外免疫途径相比，通过黏膜免疫，可以诱发IgA产生；具有免疫佐剂作用、固有免疫原性及对胆汁酸的抵抗力 It can be colonized in the respiratory system, digestive system, and urinary system, and plays an important role in maintaining the micro-ecological balance; As a food-grade bacteria, it is easy to build a food-grade gene cloning and expression system, which can effectively improve the safety of genetic engineering products, and it can simplify the post-processing process of the expression product to a certain extent; Compared with parenteral immune pathways, mucosal immunity can induce IgA production; It has immune adjuvant effect, inherent immunogenicity and resistance to bile acid	

续表

常见底盘 Common chassis	特点 Features	菌落形态 Colony morphology
枯草芽孢杆菌 *Bacillus subtilis*	遗传背景清晰，蛋白质分泌机制健全，生长迅速，易于培养，不产内毒素，遗传欠稳定，载体受体系统欠完备；适用于原核生物基因的克隆表达以及重组蛋白多肽的有效分泌 Clear genetic background, sound protein secretion mechanism, rapid growth, simple culture, no endotoxin production, genetic instability, and insufficient carrier receptor system; Suitable for the cloning and expression of prokaryotic genes and the effective secretion of recombinant protein peptides	
链霉菌 *Streptomyces*	作为抗生素、抗肿瘤活性物质、免疫抑制剂等次级代谢产物的主要来源和产生菌，决定了其在异源表达系统中无可替代的地位。其体内丰富的初级代谢途径可以为大多数次级代谢生物合成途径提供充足的前体供应；与大肠杆菌相比，链霉菌有本质的不同，比如高 GC 的基因组 DNA，以及分属革兰氏阳性菌，它们的蛋白质、启动子、增强子，以及调控元件的兼容性更优异；有其独特的转录后修饰系统 As the producing bacteria, which is the main source of antibiotics, anti-tumor active substances, immunosuppressants and other secondary metabolites, determines its irreplaceable position in the heterologous expression system. The abundant primary metabolic pathways in the body can provide sufficient precursor supply for most secondary metabolic biosynthetic pathways; Compared with *E. coli*, *Streptomyces* is essentially different, such as high GC genomic DNA and it is gram species. The compatibility of their proteins, promoters, enhancers, and regulatory elements is more outstanding; they have their unique post-transcriptional modification system	

3.2 通过代谢工程进行底盘改造
3.2 Chassis Modification Through Metabolic Engineering

图 2-3-1 所示为代谢工程中细菌底盘开发示意图，该图展示了驯化潜在的目标野生型菌株所需的关键步骤。整个过程建立在六个相互关联的主要方面，涵盖了从所关注菌株的功能基因组学和生理特性的基本认识，到专用合成生物学工具的设计和应用整个范畴。

Fig. 2-3-1 is a proposed chart for the development of a bacterial chassis for metabolic engineering, indicating the key steps required for domestication of a potentially interesting wild-type strain. The entire process builds upon six main interconnected aspects, which cover the whole range between gaining fundamental insight into functional genomics and physiology of the strain at stake and the design and adoption of dedicated synthetic biology tools.

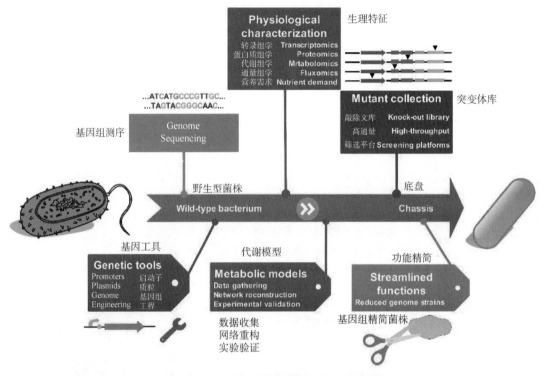

图 2-3-1 代谢工程中细菌底盘开发示意

Fig. 2-3-1 Development of a bacterial chassis for metabolic engineering chart

代谢组学是效仿基因组学和蛋白质组学的研究思想，对生物体内所有代谢物进行定量分析，并寻找代谢物与生理病理变化的相对关系的研究方式，是系统生物学的组成部分。其研究对象大都是相对分子质量为 1 000 以内的小分子物质。先进分析检测技术结合模式识别和专家系统等计算分析方法是代谢组学研究的基本方法。

Metabolomics is a research method that imitates the research ideas of genomics and proteomics, quantitatively analyzes all metabolites in the organism, and finds the relative relationship between metabolites and physiological and pathological changes. It is an integral part of systems biology. Most of its research objects are small molecular substances with a relative molecular mass of less than 1,000. Advanced analysis and detection technology combined with pattern recognition and expert systems and other computational analysis methods are the basic methods of metabolomics research.

蛋白质组学是以蛋白质组为研究对象，研究细胞、组织或生物体蛋白质组成及其变化规律的科学。蛋白质组学本质上指的是在大规模水平上研究蛋白质的特征，包括蛋白质的表达水平，翻译后的修饰，蛋白与蛋白相互作用等，由此获得蛋白质水平上的关于疾病发生、细胞代谢等过程的整体而全面的认识。

Proteomics is the science of studying the protein composition of cells, tissues or organisms and their changing laws with the proteome as the research object. Proteomics essentially refers to the study of protein characteristics at a large-scale level, including protein expression levels, post-translational modifications, protein-protein interactions, etc., so as to obtain an overall and comprehensive understanding of disease occurrence, cell metabolism and other processes at the protein level.

转录组学是一门在整体水平上研究细胞中基因转录的情况及转录调控规律的科学。转录组学是从 RNA 水平研究基因表达的情况。转录组即一个活细胞所能转录出来的所有 RNA 的总和，是研究细胞表型和功能的一个重要手段。以 DNA 为模板合成 RNA 的转录过程是基因表达的第一步，也是基因表达调控的关键环节。

Transcriptomics is a discipline that studies gene transcription in cells and the regulation of transcription at the overall level. Transcriptomics is the study of gene expression from the RNA level. The transcriptome is the sum of all RNAs that can be transcribed by a living cell. It is an important method for studying cell phenotype and function. The transcription process of synthesizing RNA using DNA as a template is the 1st step in gene expression and a key link in gene expression regulation.

3.3 微生物底盘的优化
3.3 Optimization of Microbial Chassis

图 2-3-2 所示为使用自上而下的策略设计和改造合成微生物底盘的原理。计算系统分析、实验数据和模型的出现及使用通常可以揭示细胞生存所必需的基因。随后，可以通过去除非必需基因构建合成底盘，进而在下游应用中进行验证。应用中的基因组数据可进一步促进底盘和途径的优化。

Fig. 2-3-2 shows schematic illustration of engineering and modification of synthetic microbial chassis using a top-down strategy. The advent and use of computational systems analysis and experimental data and models can often reveal genes that areessential for cellular life. Subsequently, synthetic chassis can be generated by removing non-essential genes and then be verified in downstream applications. Genomic data in applications can be of further benefit for optimizing chassis and pathways.

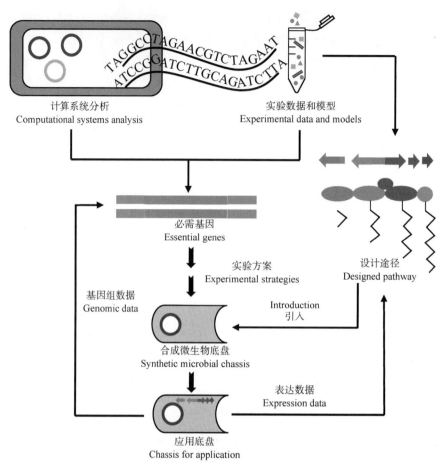

图 2-3-2 使用自上而下的策略设计和改造合成微生物底盘的原理

Fig. 2-3-2 Schematic illustration of engineering and modification of synthetic microbial chassis using a top-down strategy

自下而上的策略构建微生物底盘如图 2-3-3 所示。第一、二行显示了在大肠杆菌和酵母中进行合成和克隆来构建基因组，以及通过基因组移植测试活力。第三行显示了具备所需表型的基因、途径或基因组的进一步设计，以及使用相同的方法构建最佳合成底盘。

Construction of synthetic microbial chassis by a bottom-up strategy is shown in Fig. 2-3-3. The 1st and 2nd rows show genome building by means of synthesis and cloning in *E. coli* and yeast and testing for viability by means of genome transplantation. The 3rd row shows the further design of genes, pathways or genomes with a desired phenotype, followed by the use of the same methods to construct optimal synthetic chassis.

高通量筛选：以分子水平和细胞水平的实验方法为基础，以微板作为实验工具，以自动化操作系统执行试验过程，以灵敏、快速的检测仪器采集实验结果数据，同时通过计算机分析和处理实验数据。该技术体系可检测数以千万的样品并支持对相应数据库的操作，具有微量、快速、灵敏和准确等特点。

High throughput screening: based on experimental methods at the molecular and cellular levels, microplates are used as the experimental tool, the test process is executed with an automated operating system, the experimental result data is collected with sensitive and fast detection instruments, and the experimental data is analyzed and processed by a computer at the same time. The technical system that detects tens of millions of samples and supports operation with the corresponding database obtained, it has the characteristics of trace, fast, sensitive and accurate.

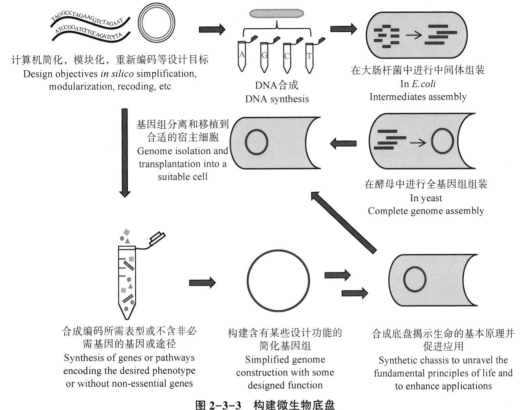

图 2-3-3 构建微生物底盘

Fig. 2-3-3 Construction of synthetic microbial chassis

3.4 理想细菌底盘的特征
3.4 Characteristics of Ideal Bacterial Chassis

图 2-3-4 所示为基于假单胞菌的理想物理和代谢合成生物学底盘的特征，该图描绘了从土壤中分离出的假单胞菌进入工业生产的过程，并总结了在基因组、代谢和生理不同细胞组织层次上的底盘所应具备的主要特征。没有任何天然存在的微生物符合理想底盘的所有要求，但合成生物学和理性代谢工程有助于满足这些要求。应注意的是，达到此目标所需的操作往往需要循环执行。

Fig. 2-3-4 shows the desirable characteristics in an ideal physical and metabolic synthetic biology chassis based on *Pseudomonas*. The figure sketches the journey of a *Pseudomonas* isolate from soil up to industrial production and summarizes the main traits needed in a chassis at the different levels of cellular organization: genomic, metabolic, and physiological. No naturally-occurring microorganism conforms with all the requirements in such ideal chassis, and synthetic biology and rational metabolic engineering help fulfilling these requirements. Note the cyclic nature of the manipulations needed to reach such objective.

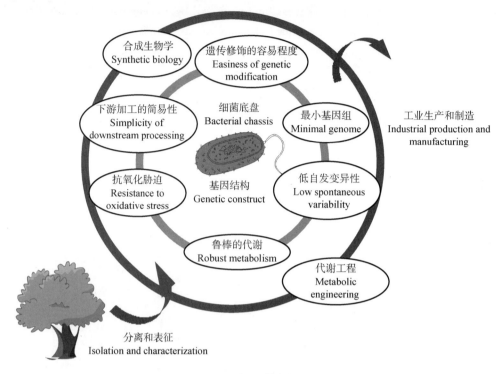

图 2-3-4 理想细菌底盘的特征
Fig. 2-3-4 Characteristics of ideal bacterial chassis

3.5 提高产量的改造策略
3.5 Engineering Strategies to Improve Production

图 2-3-5 中列出了几种针对类异戊二烯生物合成微生物宿主的改造策略，包括前体化合物支持（图 2-3-5 (a)）、辅因子支持（图 2-3-5 (b)）、阻断竞争途径（图 2-3-5 (c)）、细胞毒性工程（图 2-3-5 (d)）和微生物宿主进化（图 2-3-5 (e)）。

In Fig 2-3-5, several engineering strategies targeting microbial hosts for isoprenoid biosynthesis, including precursor support (Fig. 2-3-5 (a)), cofactor support (Fig. 2-3-5 (b)), blocking the competitive pathway (Fig. 2-3-5 (c)), cytotoxicity engineering (Fig. 2-3-5 (d)) and microbial host evolution (Fig. 2-3-5 (e)) are illustrated.

除了改造 MEP、MVA 途径和下游类异戊二烯途径外，针对中心碳代谢途径的改造策略包括前体化合物支持和辅因子支持，已得到广泛开发。当中心碳代谢途径被修改时，微生物宿主被工程化。除了前体化合物和辅因子支持外，微生物宿主还能在其他方面影响合成途径。首先，天然竞争途径可能会消耗碳流或抑制目标产物的代谢途径。其次，途径的中间体和产物可能对微生物细胞产生毒害。因此，阻断竞争途径和细胞毒性工程应运而生。

In addition to the MEP and MVA pathway engineering and the downstream isoprenoid pathway engineering, engineering strategies targeting the central carbon pathway, including precursor support and cofactor support, have been widely developed. The microbial host is engineered when the central carbon pathway is modified. In addition to precursor support and cofactor support, the microbial host can affect the synthetic pathway in other aspects. First, the native competitive pathway might consume the carbon flux or inhibit the metabolic pathway for target production. Second, the intermediates and products of the pathway might be toxic to microbial cells. Therefore, blocking the competitive pathway and cytotoxicity engineering are presented.

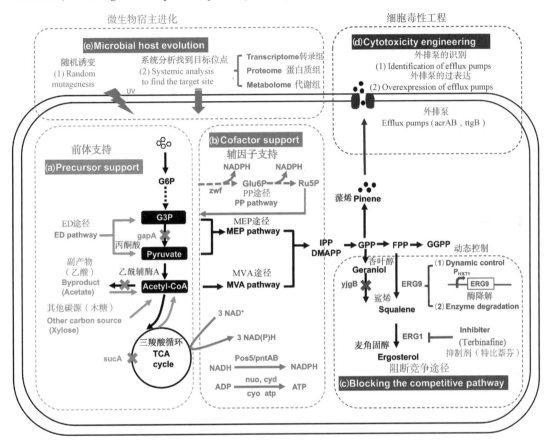

图 2-3-5　微生物宿主的工程改造策略

Fig. 2-3-5　Engineering strategies targeting microbial hosts

(a) 前体化合物支持；(b) 辅因子支持；(c) 阻断竞争途径；(d) 细胞毒性工程；(e) 微生物宿主进化

(a) Precursor support; (b) Cofactor support; (c) Blocking the competitive pathway;
(d) Cytotoxicity engineering; (e) Microbial host evolution

ED 途径又称 2-酮-3-脱氧-6-磷酸葡糖酸（KDPG）途径。因最初由 Nathan Entner 和 Michael Doudoroff 两人（1952）在嗜糖假单胞菌中发现，故命为此名。作为替代，该途径存在于某些缺乏完整 EMP 途径的微生物中，是微生物所特有的途径，特点是葡萄糖只经过 4 步反应即可快速获得由 EMP 途径经 10 步反应才能形成的丙酮酸。

ED pathway is also known as the 2-keto-3-deoxy-6-phosphate gluconic acid (KDPG) pathway. It was named because it was first discovered in *Pseudomonas saccharophila* by Nathan Entner and Michael Doudoroff (1952). This is an alternative pathway that exists in some microorganisms that lack a complete EMP pathway and it is specific to these microorganisms. The characteristic is that glucose only goes through 4 steps to quickly obtain pyruvate which can be formed by EMP pathway after 10 steps.

细胞毒性：由细胞或化学物质引起的细胞杀伤事件，不依赖于凋亡或坏死的细胞死亡机理。有时需要对特定物质进行细胞毒性检测，如药物筛选。

Cytotoxic: a cell killing event caused by a cell or chemical, independent of the mechanism of cell death by apoptosis or necrosis. Sometimes cytotoxicity tests are performed on specific substances, such as drug screening.

3.6 新型生物活性化合物的挖掘
3.6 Mining of New Bioactive Compounds

图 2-3-6 所示的流程图描述了从极端环境真菌中寻找新生物活性化合物不同方法的主要步骤。该图显示了经典方法（绿色面板）和基因组挖掘（紫色面板），以及宏基因组（红色面板）法。

The flowchart in Fig. 2-3-6 describe the main steps in different approaches to search for new bioactive compounds from fungi in extreme environments. Classical methodologies (green panels) are shown along with genome mining (purple panels) and metagenomic (red panels) approaches.

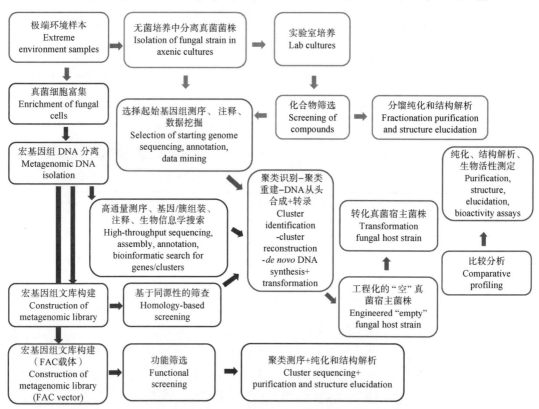

图 2-3-6 从极端环境真菌中寻找新生物活性化合物不同方法的步骤（见彩插）

Fig. 2-3-6 Main steps in different approaches to search for new bioactive compounds from fungi in extreme enrironments (see the color figure)

富集培养：利用不同微生物间生命活动特点的不同，人为地提供一些特定的环境条件，使特定种（类）的微生物旺盛生长并占据数量优势，更利于分离出特定的微生物从而进行纯培养。

Enrichment culture: using the differences in the life activity characteristics of different microorganisms, artificially provide some specific environmental conditions to make specific species (types) of microorganisms grow vigorously, make them dominant in number, and be more conducive to the separation of specific microorganisms, leading to pure culture.

广义的宏基因组是指特定环境下所有生物遗传物质的总和，它决定了生物群体的生命现象。它是以生态环境中全部DNA作为研究对象，通过克隆、异源表达来筛选有用基因及其产物，研究其功能和彼此之间的关系和相互作用，并揭示其规律的一门科学。

狭义的宏基因组学则以生态环境中全部细菌和真菌基因组DNA作为研究对象，而非采用一种传统可培养微生物的基因组。它包含了可培养和还不能培养的微生物的基因，通过克隆、异源表达来筛选有用基因及其产物，研究其功能和彼此之间的关系和相互作用，并揭示其规律。

Generalized metagenome refers to the sum of all biological genetic material under a particular environment, which determines the life phenomenon of a biological population. It is the science of cloning, which screens useful genes and their products through heteogous expression, to study their functions and relationships and interactions with each other, and to reveal their regulations, using the whole DNA of an ecological context as a research object.

Narrow sense metagenomics focuses on the total bacterial and fungal genomic DNA in an ecological context, instead of using the genome of a conventionally cultivated microorganism. It contains genes from both cultivable and not yet cultivable microorganisms, screens useful genes and their products by cloning, heterologous expression, studies of their functions and relationships and interactions with each other, and to reveal its law.

第 3 章 农业生物技术

Chapter 3　Agricultural Biotechnology

本章提要

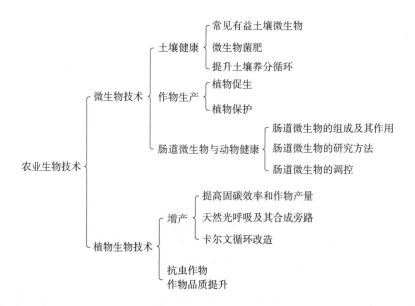

Summary of Chapter 3

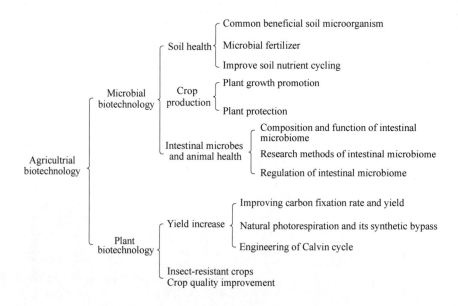

Ⅰ 微生物技术
1 Microbial Biotechnology

1.1 土壤健康
1.1 Soil Health

1.1.1 常见有益土壤微生物
1.1.1 Common Beneficial Soil Microorganism

表 3-1-1 所示为常见有益土壤微生物。
Table 3-1-1 shows common beneficial soil microorganism.

表 3-1-1 常见有益土壤微生物
Table 3-1-1 Common beneficial soil microorganism

土壤有益微生物 Beneficial soil microorganisms	主要种类 Main species	对病原菌的作用方式 Mode of action on pathogens	形态 Morphology
细菌 Bacteria	芽孢杆菌、假单胞菌、土壤农杆菌 *Bacillus*, *pseudomonas*, *agrobacterium*	抗生作用、竞争、寄生作用和胞外酶的生产、诱导抗性 Antibiotic effect, competition, parasitic effect and production of extracellular enzymes, induced resistance	芽孢杆菌 *Baillus*
真菌 Fungus	木霉菌、毛壳菌、厚壁孢子轮枝菌 *Trichoderma*, *chaetomium*, *verticillium chlamydosporium*		木霉菌 *Trichoderma*
放线菌 Actinomycetes	链霉菌属 *Streptomyces*		链霉菌 *Streptomyces*

1.1.2 微生物菌肥
1.1.2 Microbial Fertilizer

表 3-1-2 所示为微生物菌肥种类及其作用。

Table 3-1-2 shows types of functions of microbial fertilizer.

表 3-1-2 微生物菌肥种类及其作用

Table 3-1-2 Types and functions of microbial fertilizer

按微生物种类分类 Classified by microorganism	按功能和肥效分类 Classified by function and fertilizer	作用 Effects	形态 Morphology
细菌类肥料（根瘤菌肥、固氮菌肥、解磷菌肥、解钾菌肥、光合菌肥） Bacterial fertilizers (rhizobium fertilizer, nitrogen-fixing bacteria fertilizer, phosphate solubilizing bacteria fertilizer, photosynthetic bacteria fertilizer)	增加氮素营养的菌肥（根瘤菌肥、固氮菌肥等） Increase soil nitrogen nutrient bacterial fertilizer (rhizobium fertilizer, nitrogen-fixing fertilizer, ect.)	增进土壤肥力（固氮微生物肥料可以增加土壤中的氮素含量；溶磷解钾的微生物可增加适宜作物吸收的磷、钾化合物） Increase soil fertility (nitrogen-fixing microbial fertilizers increase the total nitrogen in the soil; microorganisms that dissolve phosphorus and potassium can increase the phosphorus and potassium for crop absorption)	根瘤菌 *Rhizobium*
真菌肥料（菌根真菌、外生菌根菌剂和内生菌根菌剂） Fungal fertilizers (mycorrhizal fungal fertilizer, ectomycorrhizal inoculants and endomycorrhizal inoculants)	分解土壤难溶性矿物质的菌肥（解磷细菌肥、解钾解细菌菌肥、菌根真菌肥） Microbial fertilizer that decomposes insoluble minerals in the soil (phosphate solubilizing bacterial fertilizer, potassium solubilizing bacterial fertilizer, mycorrhizal fungal fertilizer)	改良土壤结构（分泌胞外多糖物质，可参与腐殖质形成，改良土壤理化性质） Improve soil structure (secreted extracellular polysaccharides can participate in the formation of humus and improve soil physical and chemical properties)	木霉菌 *Trichoderma*

1.1.3 提升土壤养分循环
1.1.3 Improve Soil Nutrient Cycling

在图 3-1-1 中，分解、同化和非生物过程用黑色箭头表示，异化过程用其他颜色箭头表示：固氮作用（灰色）、硝化作用（浅蓝色）、硝化细菌硝化（深蓝色）、硝化细菌反硝化（浅绿色）、反硝化（深绿色）、非反硝化氧化亚氮还原（橄榄绿色）、硝酸铵异化还原（黄色）和氨厌氧氨氧化（紫色）。DNRA，硝酸铵异化还原；SON，土壤有机氮；DON，可溶性有机氮；GHG，温室气体。带有星（*）号的过程在热带森林土壤中的发生和生态学现象尚未得到研究。

In Fig. 3-1-1, the arrows of decomposition, assimilative and abiotic processes are shown in black and the arrows of dissimilative processes are shown in different colors: N-fixation (gray), nitrification (light blue), nitrifier nitrification (dark blue), nitrifier denitrification (light green), denitrification (dark green), non-denitrification N_2O reduction (olive green), DNRA (yellow), and anammox (purple). DNRA, dissimilatory nitrate reduction to ammonium; SON, soil organic N; DON, dissolved organic N; GHG, greenhouse gas. Occurrence and ecology of these processes with a star (*) in tropical forest soils has not been yet investigated.

生物固氮：指固氮微生物将大气中的氮气还原成氨的过程。

Biological nitrogen fixation: a microbially mediated process that converts nitrogen gas to ammonia.

硝化作用：氨在有氧的条件下，经亚硝酸细菌和硝酸细菌的作用转化为硝酸的过程。

Nitrification: a microbial process which ammonia is aerobically oxidized to nitrate by nitrite bacteria and nitrate bacteria.

反硝化作用：微生物将硝酸盐及亚硝酸盐还原为气态氮化物和氮气的过程。

Denitrification: a microbial process of reducing nitrate and nitrite to gaseous forms of nitride and nitrogen.

矿化作用：在微生物作用下，有机态化合物转化为无机态化合物过程的总称。

Mineralization: the microbial process by which chemicals present in organic matters are decomposed into inorganic matters.

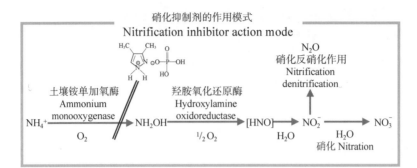

图 3-1-1 热带森林土壤生物氮循环的示意图（见彩插）

Fig. 3-1-1 Schematic representation of the biological N cycle in tropical forest soils（see the color figure）

1.2 作物生产
1.2 Crop Production

1.2.1 植物促生
1.2.1 Plant Growth Promotion

A. 植物根际微生物的种内、种间关系及对不同环境的应对机制

A. Intraspecific and interspecific relationships of plant rhizospheric microorganisms and counter measures in different environments

图 3-1-2 所示为植物根际微生物的种内、种间关系及对不同环境的应对机制。

Fig. 3-1-2 shows intraspecific and interspecific relationships of plant rhizospheric microorganisms and counter measures in different environments.

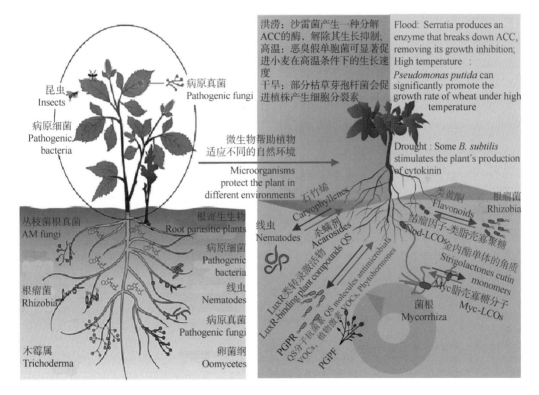

图 3-1-2 植物根际微生物的种内、种间关系及对不同环境的应对机制

Fig. 3-1-2 Intraspecific and interspecific relationships of plant rhizospheric microorganisms and counter measures in different environments

植物根际促生细菌：指生存在植物根圈范围中，对植物生长有促进或对病原菌有拮抗作用的有益细菌。

Plant growth promoting rhizobacteria (PGPR): bacteria associated with plant roots that promote plant growth and antagonize pathogenic microorganisms.

植物促生真菌（PGPF）：指生存在植物根圈范围中，对植物生长有促进或对病原菌有拮抗作用的有益真菌的统称。

Plant growth-promoting fungi（PGPF）：beneficial fungi associated with plant roots that promote plant growth and antagonize pathogenic microorganisms.

B. 植物根际促生细菌诱导植物抗旱性的机理

B. Mechanisms of plant drought tolerance induced by plant growth-promoting rhizobacteria

植物根际促生细菌诱导植物抗旱性的机理如图 3-1-3 所示。

Fig. 3-1-3 shows mechanism of plant drought tolerance induced by plant growth-promoting rhizobacteria.

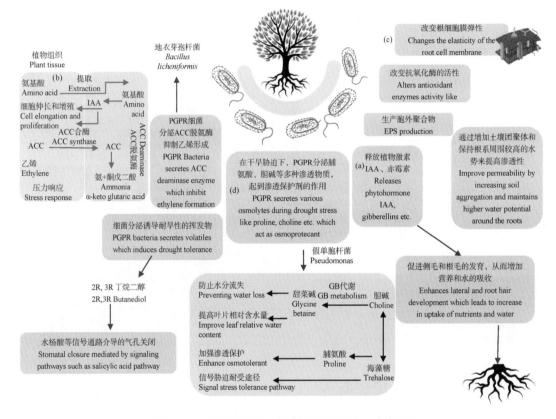

图 3-1-3　植物根际促生细菌诱导植物抗旱性的机理

Fig. 3-1-3　Mechanism of plant drought tolerance induced by plant growth-promoting rhizobacteria

植物根际促生细菌在干旱胁迫下提高植物生长能力的机理如下：ⓐ产生不同类型的激素，如生长素、吲哚乙酸、脱落酸、赤霉素和细胞分裂素；ⓑ产生氨基环丙烷羧酸脱氨酶，抑制了乙烯形成途径，降低根部乙烯水平；ⓒ诱导系统抗性（ISR）；ⓓ产生多种渗透物。

The plant growth-promoting rhizobacteria improve plant growth and performance under drought stress by the following mechanisms：ⓐproduce different types of phytohormones, such as auxin, IAA, ABA, gibberellic acid and cytokinins，ⓑproduce ACC deaminase, which inhibits the ethylene formation pathway to decrease the ethylene level in the roots，ⓒinduced systemic resistance

(ISR), produce various osmolytes.

1.2.2 植物保护
1.2.2 Plant Protection

A. 生物农药发展

A. Development of biopesticides

生物农药：以生物活体（真菌，细菌，昆虫病毒，转基因生物，天敌等）或其代谢产物（信息素，生长素，萘乙酸，2，4-D 等）制成的杀灭或抑制农业有害生物的制剂。生物农药的发展时间线如图 3-1-4 所示。

Biopesticide: preparations that consist of living organisms (fungi, bacteria, insect viruses, genetically modified organisms, natural enemies, etc.) or their metabolites (pheromone, auxin, naphthaleneacetic acid, 2, 4-D, etc.) to kill or inhibit agricultural pests. Fig. 3-1-4 shows timeline of biopesticides.

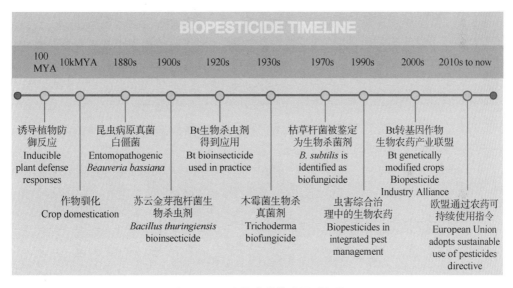

图 3-1-4 生物农药的发展时间线

Fig. 3-1-4 Timeline of biopesticides

B. 苏云金芽孢杆菌的功效

B. Potentials of bacillus thuringiensis

图 3-1-1 (a) 所示为 Bt 细胞工厂的功效；图 3-1-5 (b) 所示为 Bt 对线虫的作用方式。

Fig. 3-1-5 (a) shows Bt cell factory potentials; Fig. 3-1-5 (b) shows action mode of Bt against nematodes.

晶体蛋白在孢子萌发后破坏肠道，Bt 细胞生长过程中产生的几丁质酶、金属蛋白酶、外毒素等多种致病因子与晶体蛋白质具有协同作用。

The crystal proteins destroy the intestine following spore germination. Multi-pathogenic factors produced during Bt cell growth, such as chitinase, metalloproteinase and exotoxin, can act synergistically with crystal proteins.

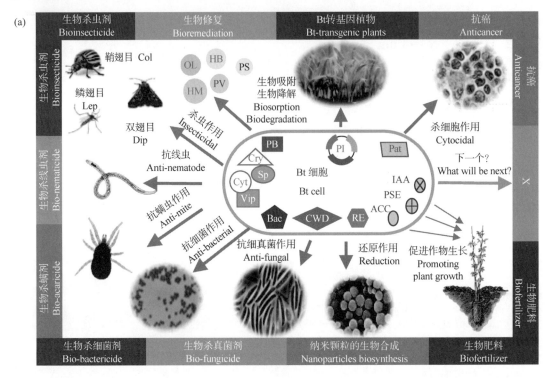

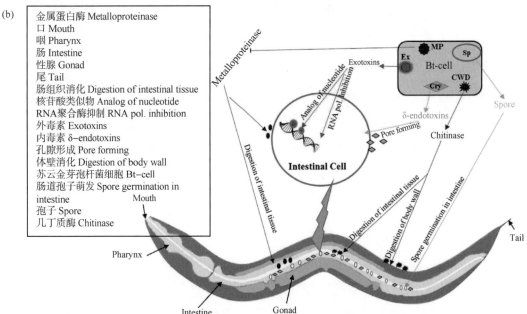

图 3-1-5 苏云金芽孢杆菌的功效及其对线虫的杀灭方式
Fig. 3-1-5 Potentials of *bacillus thuringiensis* and its pesticidal effects on nematodes
(a) Bt 细胞工厂的功效;(b) Bt 对线虫的作用方式
(a) Bt cell factory potentials; (b) Action mode of Bt against nematodes

注释：
ACC，氨基环丙烷羧酸脱氨酶；Bac，细菌素；CWD，细胞壁降解酶；Col，鞘翅目；Cry，晶体蛋白（δ-内毒素）；Cyt，细胞溶解蛋白质；Dip，双翅目；DY，染料；HB，除草剂；HM，重金属；IAA，吲乙酸；Lep，鳞翅目；OL，油（石油）；Par，伴孢；PB，生物修复相关蛋白质；Pl，质粒；PS，生物农药；PSE，磷酸溶解酶；PV，塑料；RE，降解酶；Sp，孢子；Vip，植物杀虫蛋白。

Note：
ACC, 1-aminocyclopropane-1-carboxylic acid deaminase; Bac, bacteriocin; CWD, cell wall degrading enzymes; Col, Coleoptera; Cry, crystal proteins (δ-endotoxins); Cyt, cytolytic proteins; Dip, Diptera; DY, dyes; HB, herbicides; HM, heavy metals; IAA, indole-3-acetic acid; Lep, Lepidoptera; OL, oil (petroleum); Par, parasporin; PB, bioremediation involving proteins; Pl, plasmid; PS, pesticide; PSE, phosphate solubilization enzymes; PV, plastics; RE, reducing enzymes; Sp, spore; Vip, vegetative insecticidal proteins.

苏云金芽孢杆菌伴孢晶体：苏云金芽孢杆菌在形成芽孢的同时，会在芽孢旁形成一颗菱形、方形或不规则形状的碱溶性蛋白质晶体，称为伴孢晶体。

Bacillus thuringiensis parasporal crystal：diamond-shaped, square-shaped or irregular-shaped alkali-soluble protein crystal produced by *Bacillus thuringiensis* when forming spores.

诱导系统抗性（ISR）：在植物根际附近（包括根围和根内）存在大量的细菌，其中非致病性根际细菌能诱导植物产生系统抗性，称之为 ISR，其表型与病菌诱导的系统获得性抗性（SAR）相似。

Induced system resistance (ISR)：a resistance mechanism in plants that is induced by non-pathogenic bacteria living around or inside plant root, which is similar to the pathogen-induced systemic acquired resistance.

生物修复：利用生物的生命代谢活动减少污染环境中有毒有害物的浓度或使其无害化，从而使污染了的环境能够部分地或完全地恢复到原初状态的过程。

Bioremediation：the process of using biological metabolism to reduce the concentration of toxic and harmful substances in the polluted environment or make them harmless, so that the polluted environment can be partially or completely restored to its original state.

C. 阿维菌素的合成

C. Biosynthesis of avermectin

阿维菌素的生物合成如图 3-1-6 所示。

Fig. 3-1-6 shows biosynthesis of avermectin.

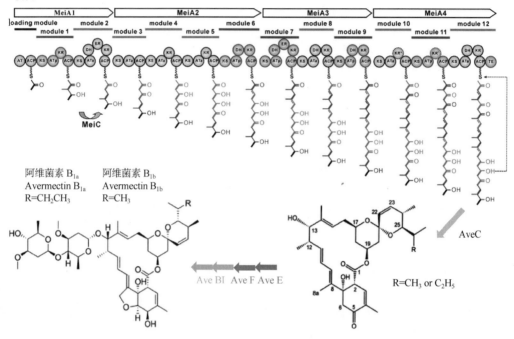

图 3-1-6 阿维菌素的生物合成

Fig. 3-1-6 Biosynthesis of avermectin

D. 生物制药

D. Bio-pharmaceuticals

识别新药物分子的主要策略如图 3-1-7 所示。

Fig. 3-1-7 shows main strategies to identify new biopharmaceutical molecules.

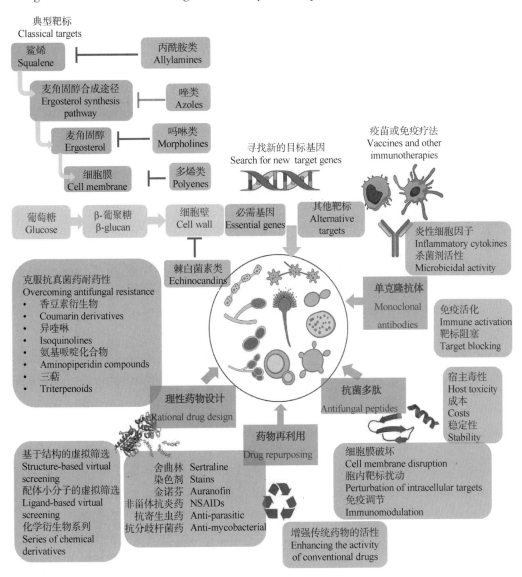

图 3-1-7　识别新药物分子的主要策略

Fig. 3-1-7　Main strategies to identify new biopharmaceutical molecules

1.3 肠道微生物与动物健康
1.3 Intestinal Microbes and Animal Health

1.3.1 肠道微生物的组成及其作用
1.3.1 Compostion and Function of Intestinal Microbiome

A. 动物肠道微生物区系组成
A. Composition of animal intestinal microflora

肠道菌群是指生活在胃肠道特定环境中的微生物群落，包括细菌、古生菌、原生动物和瘤胃真菌。动物肠道微生物区系组成如图3-1-8所示。

The gut microbiota refers to the community of microorganisms inhabiting a defined environment along the gastrointestinal tract, and includes bacteria, fungi, protozoa, and archaea. Fig. 3-1-8 shows composition of microflora of animal gut.

菌种 Microorganisms	瘤胃微生物 Rumen microorganisms/mL	举例 Examples	菌体形态 Morphology
细菌 Bacteria	$10^{10} \sim 10^{11}$	丙酸菌属、双歧杆菌属、毛螺菌属、瘤胃球菌属、梭菌属、消化链球菌属、纤维杆菌、反刍杆菌、普氏菌属、琥珀酸弧菌属、硒单胞菌 *Propionibacterium, Bifidobacterium, Lachnospira, Ruminococcus, Clostridium, Peptostreptococcus, Fibrobacter, Ruminobacter, Prevotella, Succinivibrio, Selenomonas*	 链球菌属 *Streptococcus* 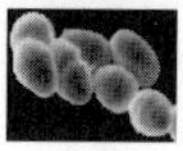 瘤胃球菌属 *Ruminococcus*
古生菌 Archaea	$10^{8} \sim 10^{9}$	梭菌属、甲烷杆菌、瘤胃甲烷短杆菌、运动甲烷微菌 *Fusobacterium, Methanobacterium bryantii, Methanobrevibacter ruminantium, Methanomicrobium mobile*	梭菌属 *Fusobacterium*  甲烷杆菌属 *Methanobacterium*
原生生物 Protozoa	$10^{5} \sim 10^{6}$	全毛亚纲（肠等毛虫、多毛属），鞭毛虫类（毛滴虫属、唇鞭毛虫属） *Holotrichs (Isotricha, Dasytricha), flagellates (Trichomonas, Chilomastix).*	 鞭毛虫 *Flagellates*
瘤胃真菌 Rumen fungi	$10^{3} \sim 10^{4}$	新美鞭菌属、梨囊鞭菌属、盲肠鞭菌属、根囊鞭菌属、厌氧鞭菌属 *Neocallimastix, Piromyces, Caecomyces, Orpinomyces, Anaeromyces*	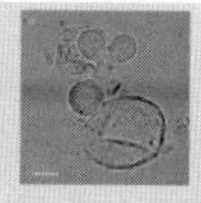 盲肠鞭菌属 *Caecomyces*

图 3-1-8 动物肠道微生物区系组成
Fig. 3-1-8 Composition of microflora of animal gut

B. 反刍动物消化道及微生物群

B. Gastrointestinal tract of ruminant animal and microbiota

图 3-1-9（a）所示为反刍动物胃肠道示意图；图 3-1-9（b）所示为瘤胃壁结构和微生物多样性及其功能。细菌是瘤胃中数量最多的微生物群，瘤胃液中的菌体密度为 $10^{10} \sim 10^{11}$ 个细胞/mL。瘤胃细菌群以厚壁菌门、拟杆菌门和变形菌门为主，包含 *Prevotella*，*Fibrobacter* 和 *Butyrivibrio* 等许多属，能够代谢一系列膳食多糖和肽。古生菌是瘤胃产甲烷菌（$10^6 \sim 10^8$ 个细胞/mL 瘤胃液），均属于 Euryarchaeota 门，主要由 *Methanobrevibacter ruminantium* 和 *Methanobrevibacter gottschalkii* 进化枝的成员组成。原生动物：瘤胃液纤毛虫的数量为 $10^4 \sim 10^6$ 个细胞/mL，最丰富的属是 *Entodinium*、*Polyplastron*、*Epidinium* 和 *Eudiplodinium*。厌氧真菌仅在 19 世纪 70 年代发现，数量为 $10^3 \sim 10^6$ 个游动孢子/mL 瘤胃液，瘤胃中的纤维素分解厌氧真菌属于 Neocallimastigomycota 门，目前分为 8 个属（*Neocallimastix*、*Piromyces*、*Ontomyces*、*Buwchfawromyces*、*Caecomyces*、*Orpinomyces*、*Anaeromyces* 和 *Cyllamyces*）。噬菌体/古噬菌体是瘤胃病毒组以尾病毒目为主，噬菌体是微生物种群的关键调节因素和水平基因转移的促进因素。

Fig. 3-1-9 (a) shows schematic of the ruminant animal gastrointestinal tract and Fig. 3-1-9 (b) shows depiction of rumen wall structure and microbial diversity and function. Bacteria is the most numerous microbial group in the rumen, bacteria are present at a density of $10^{10} - 10^{11}$ cells/mL rumen fluid. The rumen bacteriome is dominated by members of the Firmicutes, Bacteroidetes, and Proteobacteria phyla, containing numerous genera like *Prevotella*, *Fibrobacter* and *Butyrivibrio*, capable of metabolizing a range of dietary polysaccharides and peptides. Archaea is the rumen methanogens ($10^6 - 10^8$ cells/mL rumen fluid) which belong exclusively to the Euryarchaeota phylum and are dominated by members of the *Methanobrevibacter ruminantium* and *Methanobrevibacter gottschalkii* clades. Protozoa is the ciliates that are found in the range of $10^4 - 10^6$ cells/mL in the rumen fluid, and the most abundant genera are *Entodinium*, *Polyplastron*, *Epidinium*, and *Eudiplodinium*. Anaerobic fungi is discovered only in the 1970s and present at rates of $10^3 - 10^6$ zoospores/mL, the cellulolytic anaerobic fungi in the rumen belong to the phylum Neocallimastigomycota and are currently grouped into eight genera (*Neocallimastix*, *Piromyces*, *Ontomyces*, *Buwchfawromyces*, *Caecomyces*, *Orpinomyces*, *Anaeromyces*, and *Cyllamyces*). Bacteriophage/archaeaphage is the rumen virome which is dominated by *Caudovirales*, and the phage are key regulators of microbial populations and facilitators of horizontal gene transfer.

Abbreviation：VFA, volatile fatty acid.

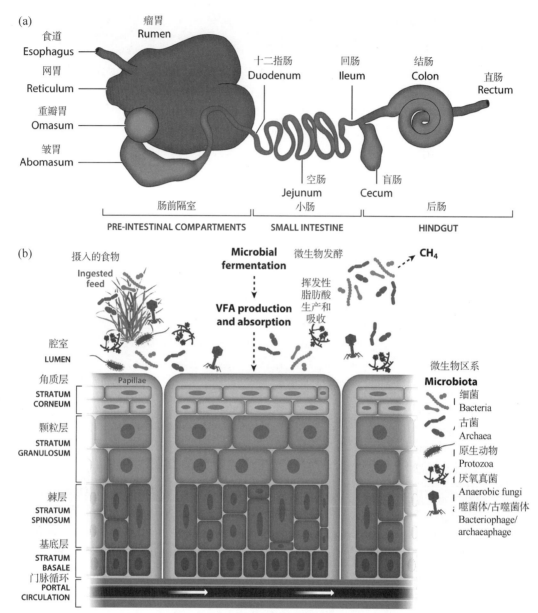

图 3-1-9　反刍动物消化道结构及其微生物区系

Fig. 3-1-9　Structure of ruminant animal gastrointestinal tract and its microbiota

(a) 反刍动物胃肠道示意图；(b) 瘤胃壁结构和微生物多样性及其功能

(a) Schematic of the ruminant animal gastrointestinal tract;
(b) Depiction of rumen wall structure and microbial diversity and function

第3章 农业生物技术

C. 益生菌及其在动物中的应用

C. Probiotics and its applications in animal health

益生菌是严格挑选的微生物活株，摄入后有益于宿主健康。益生菌对动物或人类肠道健康的功效如图 3-1-10 所示。图 3-1-11 所示为补充酵母对动物健康和营养的积极影响。

Probiotics are living microorganisms that, when ingested, provide a health benefit. Fig. 3-1-10 shows effects of probiotics on animals or human instestinal health. Fig. 3-1-11 shows the positive influence of yeast supplementation on animal health and nutrition.

- 产生具有杀菌或抑菌特性的物质
- 通过益生菌发酵活性降低腔内pH值
- 产生过氧化氢抑制革兰氏阴性菌的生长
- 影响病原微生物的代谢和毒素产生

- Producing substances that have bactericidal or bacteriostatic properties
- Decreasing luminal pH via probiotic fermentative activity
- Inhibiting the growth of Gram-negative bacteria by the hydrogen peroxide produced
- Affecting the metabolism and toxin production of the pathogenic microorganisms

- 提高消化酶的产量和活性
- 影响猪肠道的吸收和分泌活性
- 生产维生素

- Increasing digestive enzyme production and activities
- Affecting the absorption and secretion activities of swine gut
- Producing some vitamins

直接抑制作用
Direct antimicrobial inhibition

调节营养物质消化率
Modulation of nutrient digestibility

- 竞争肠壁附着部位
- 竞争肠道内的有机质或营养物质

- Competing for intestinal adhesion sites
- Competing for organic substrates or nutrients in intestinal tract

减少腹泻及抗毒素作用
Diarrhea reduction and antitoxin effects

益生菌

调节宿主免疫反应
Modulation of host immune responses

- 抑制致病菌的毒素表达
- 中和致病菌产生的肠毒素
- 通过益生菌的高发酵活性
- 提高消化酶的产量和活性
- 影响猪肠道的吸收和分泌活性

- Inhibiting toxin expression in pathogenic bacteria
- Neutralizing the enterotoxins produced by pathogenic bacteria
- By the high fermentative activity of probiotics
- Increasing digestive enzyme production and activities
- Affecting the absorption and secretion activities of swine gut

其他作用方式
Other action modes

调节肠道菌群
Modulation of gut microbiota

竞争性排斥
Competitive exclusion

- 抗氧化活性，减轻压力
- 改变细菌和宿主基因的表达

- Antioxidative activity and alleviation of stress
- Altering bacterial and host gene expression

- 通过恢复肠道屏障的完整性和功能来提高肠道先天免疫
- 通过增加肠粘液的产生或氯化物的分泌来提高肠道先天免疫
- 刺激或抑制动物的获得性免疫反应
- 通过代谢产物、细胞壁成分和DNA等影响动物免疫系统

- Improving gut innate immunity through restitution of intestinal barrier integrity and function
- Improving gut innate immunity through increasing gut mucus production or chloride secretion
- Stimulating or suppressing animal acquired immune responses
- Influencing animal immune system by metabolites, cell wall components, and DNA

图 3-1-10 益生菌对动物或人类肠道健康的功效

Fig. 3-1-10 Effects of probiotics on animals or human intestinal health

图 3-1-11 补充酵母对动物健康和营养的积极影响

Fig. 3-1-11 The positive influence of yeast supplementation on animal health and nutrition

1.3.2 肠道微生物的研究方法
1.3.2 Research Methods of Intestinal Microbiome

A. 肠道微生物重点研究领域
A. Key research areas on intestinal microbiome

肠道菌群的组学研究技术如图 3-1-12 所示。

Fig. 3-1-12 shows omics techniques for studying gut microbiota.

第3章 农业生物技术

微生物学
基于组学的方法正在取代传统的纯培养方法
Microbiological
Omics-based methods are now used as an alternative to replace traditional culture methods

宏基因组学
一种以环境样本微生物基因组为研究对象，通过筛选功能基因或测序分析为研究方法的学科
Metagenome
Macro genomics is a kind of environmental samples to the microbial genome as the research object, by screening functional genes and/or sequencing analysis as research methods

宏转录组学
微生物组水平的基因表达谱可直接反映微生物群落的活动
Metatranscriptomics
The profiling of microbiome-wide gene expression (RNA-seq) can directly inform on the activity of microbial communities

代谢组学
使用质谱或核磁共振谱测定小分子代谢物谱（代谢组）
Metabolomics
The determination of small-molecule metabolite profiles (metabolomes) can be performed using mass-spectrometry or NMR spectroscopy

鸟枪法全基因组测序
可用于识别更多种类微生物（细菌、真菌、病毒和原生生物）的整个基因组
Shotgun metagenomics sequencing
Shotgun metagenomics allows the identification of whole genomes including a larger range of micro-organisms (bacteria, fungi, viruses and protists)

16S rRNA 基因扩增子测序
16S rRNA 基因高变区域的扩增可用于测量肠道微生物的组成，方法是将扩增子聚类为操作分类单元（OTUs）的集合中，或通过分析精确序列变异（ESV）
16S rRNA gene amplicon sequencing
The amplification of 16S rRNA gene hypervariable regions can be used to measure the composition of the gut microbiome by cluttering amplicon reads into bins called operational taxonomic units (OTUs), or by determining exact sequence variants (ESV)

宏蛋白质组学
研究所有直接从环境来源中回收的蛋白质样本，可以使用基于凝胶的蛋白质组学方法，也可以使用鸟枪法
Metaproteomics
Metaproteomics is the study of all protein samples recovered directly from environmental sources, either using gel-based proteomics approaches, or directly by shotgun proteomics

图 3-1-12 肠道菌群的组学研究技术

Fig. 3-1-12 Omics techniques for studying gut microbiota

B. 微生物宏基因组研究方法

B. Research methods for microbial metagenomics

图 3-1-13 所示为宏基因组学研究流程。步骤 1：研究设计和实验方案。该步骤的重要性在宏基因组学中经常被低估；步骤 2：计算预处理。计算质量控制（QC）步骤可最大限度地减少基础序列偏差或人为因素，如去除测序接头、质量修整、去除重复序列（使用 FastQC、Trimmomatic121 或 Picard 工具）。外来或非靶标 DNA 序列也会被过滤，如需比较分类多样性或功能多样性，样本会被二次采样以对读段数量进行标准化。步骤 3：序列分析。该步骤应包括"基于读段"和"基于组装"的方法组合，具体取决于实验目标。两种方法各有优点和局限性。步骤 4：后处理。可以使用各种多元统计技术来解释数据。步骤 5：检验。由高维生物数据得到的结论容易受研究驱动偏差的影响，后续分析至关重要。

Fig. 3-1-13 shows summary of a metagenomics workflow. Step 1: study design and experimental protocol. The importance of this step is under estimated in metagenomics. Step 2: computational pre-processing. Computational quality control (QC) steps minimize fundamental sequence biases or artifacts such as removal of sequencing adaptors, quality trimming, removal of sequencing duplicates (using FastQC, Trimmomatic121 or Picard tools). Foreign or non-target DNA sequences are also filtered, and samples are subsampled to normalize read numbers if the diversity of taxa or functions is compared. Step 3: sequence analysis. This should comprise a combination of "read-based" and "assembly-based" approaches depending on the experimental objectives. Both approaches have advantages and limitations. Step 4: post-processing. Various multivariate statistical techniques can be used to interpret the data. Step 5: validation. Conclusions from high-dimensional biological data are susceptible to study-driven biases, so follow-up analyses are vital.

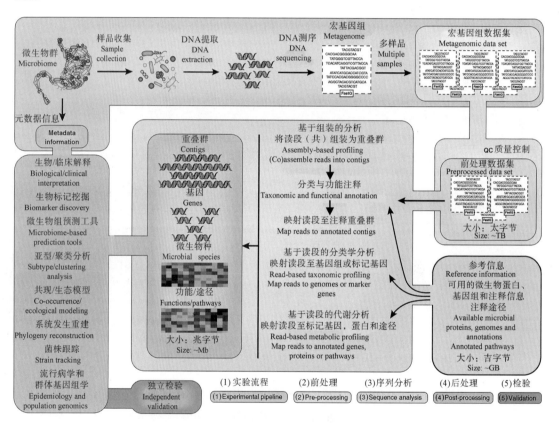

图 3-1-13 宏基因组学研究流程

Fig. 3-1-13　Summary of a metagenomics workflow

C. 蛋白质组学和宏蛋白质组学实验流程

C. Proteomics and macroproteomics laboratory procedures

图 3-1-14 所示为蛋白质组学和宏蛋白质组学实验流程。收集小鼠幼仔的整个肠道，与粪便一起制备成匀浆。经蛋白质提取和胰蛋白酶消化后，使用数据依赖性采集技术（DDA），即信息依赖性采集技术（IDA），和数据独立性采集技术（DIA、SWATH）获得 LC-MS/MS 数据，分别用于蛋白质鉴定和定量。采用运行 IDA 的数据库检索结果，以 OneOmics 云应用程序进行 SWATH 分析。对于细菌蛋白，可通过 BLAST+检索每个样品的高置信度多肽，生成的 xml 文件可导入 MEGAN 软件进行分类和功能分析。

Fig. 3-1-14 shows proteomics and metaproteomics experimental workflow. The whole intestinal tract from mice pups was collected and homogenized together with faeces. After protein extraction and tryptic digestion, LC-MS/MS data were acquired in the data-dependent acquisition (DDA), also known as information-dependent acquisition (IDA), and data-independent acquisition (DIA, SWATH) modality, for protein identification and quantitation respectively. The OneOmics cloud applications were employed for the SWATH analysis, using libraries coming from database searches on the IDA runs. For bacterial proteins, taxonomic and functional analysis were performed with MEGAN, importing the xml output obtained by a BLAST+ search on high confidence peptide list for each sample.

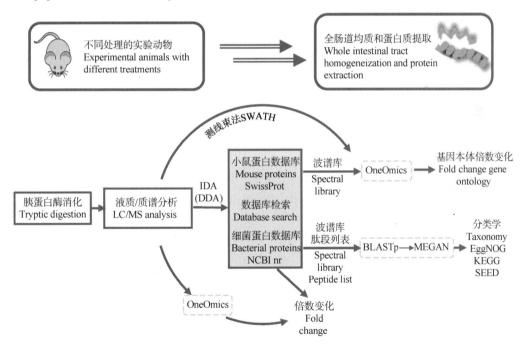

图 3-1-14　蛋白质组学和宏蛋白质组学实验流程

Fig. 3-1-14　Proteomics and metaproteomics experimental workflow

1.3.3 肠道微生物的调控
1.3.3 Regulation of Intestinal Microbiome

如图 3-1-15 所示，大量纤维组分不仅在人和其他单胃动物的大肠中发酵，而且也在单胃动物的小肠中降解。本研究表明不同膳食纤维可改变猪模型中微生物群落和短链脂肪酸的合成，其中乳酸主要在前肠中产生，丙酸盐和丁酸盐主要在后肠中产生。

As shown in Fig. 3-1-15, substantial fiber fractions not only were fermented in the hindgut of human and other monogastric animals, but also were degraded in the foregut of monogastric animals. This study suggested that different dietary fibers could alter the microbial community and SCFAs production. Lactate was mainly produced in the foregut, and propionate and butyrate were more produced in the hindgut.

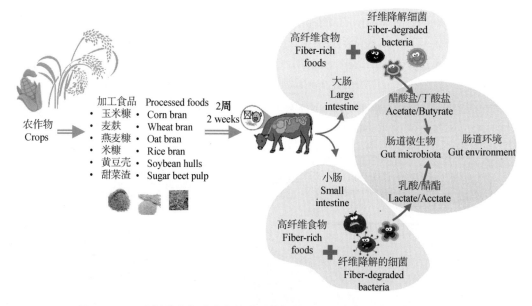

图 3-1-15 高纤维食物对动物肠道细菌群落和短链脂肪酸产生的影响
Fig. 3-1-15 Fiber-rich foods affected gut bacterial community and short-chain fatty acids production

图 3-1-16 所示为参与短链脂肪酸代谢的肠道细菌的功能和系统发育群。

Fig. 3-1-16 shows functional and phylogenetic groups of gut bacteria involved in the metabolism of short-chain fatty acids.

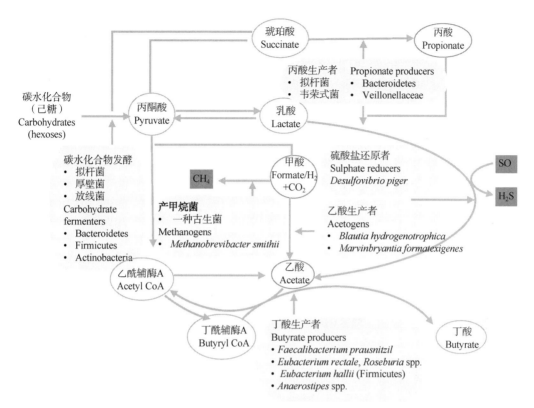

图 3-1-16　参与短链脂肪酸代谢的肠道细菌的功能和系统发育群
Fig. 3-1-16　Functional and phylogenetic groups of gut bacteria involved in the metabolism of short-chain fatty acids

粪菌移植（FMT）过程及其应用如图 3-1-17 所示。将健康供体粪便中的功能菌群移植到患者胃肠道内，重建新的肠道菌群或增加有益微生物，实现肠道及肠道外疾病的治疗。选择合适的供体是粪菌移植的关键。配偶或近亲历来被认为是理想的 FMT 捐赠者。由于环境因素相同，配偶的粪便可能会将感染传播的风险降到最低。因近亲间的微生物物种受体相似，黏膜免疫系统中的适应性免疫可能对来自亲属的微生物类群表现出更强的耐受性。尽管如此，更多的临床证据已经证明供体选择和粪菌移植成功率之间并不存在联系。新鲜粪便材料应在供体产生后 6 h 内进行处理，并可在室温下保存，直至进一步处理。目前，粪菌移植的方式包括上消化道途径、下消化道途径（通过结肠镜检查或保留灌肠）和口服胶囊。FMT 手术的实际准备，包括标准的肠道准备与其他内镜手术类似。为了保证移植体的健康，在移植之前，肠内应当几乎没有受污染的粪便物质。

Fecal microbiota transplantation (FMT): process and its application is shown in Fig. 3-1-17. Transferring fecal bacteria and other microbes from a healthy individual into a patient individual to try to reintroduce or boost helpful organisms. FMT has been used to treat gastrointestinal and other diseases. The selection of appropriate donors is the key to FMT. Spouses or close relatives have traditionally been considered as ideal FMT donors. Feces from spouses may minimize the risk of infection transmission because there are common environmental factors. The expected receptor is similar to its close relatives in microbial species. Therefore, adaptive immunity in mucosal immune system may show stronger tolerance to donor microbiota. However, more clinical evidences have proved that there is no link between donors and FMT results. Fresh fecal material should be processed within 6 h after donor production and can be stored at room temperature until further processing. At present, the management of fecal materials includes upper gastrointestinal pathway, lower gastrointestinal pathway (through colonoscopy or retention enema) and oral capsule. The actual preparation for FMT surgery is similar to any other endoscopic surgery, including standard intestinal preparation. In order to ensure the health of the graft, there should be almost no contaminated fecal material in the intestine before fecal infusion.

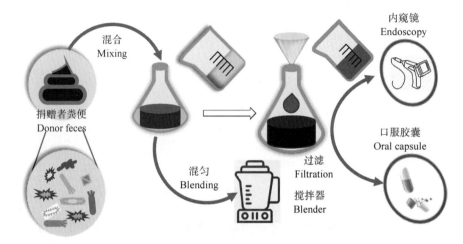

图 3-1-17 粪菌移植过程及其应用

Fig. 3-1-17 Fecal bacteria transplantation process and its application

FMT 在动物中的应用可分为三个方向，即①治疗性应用、②预防性应用、③刺激病原特异性免疫。

The use of FMT in animals can be divided into three potential applications, including ①therapeutic application, ② prophylactic application, and ③ application of FMT for stimulating pathogen-specific immunity.

当给药的目的是治疗临床症状或解决正在发作的疾病时使用 FMT 治疗。相反，FMT 预防是在暴露于病原体之前提供有益的微生物群。FMT 还可被用作类似于疫苗接种的免疫刺激工具，其中移植材料可激发病原体特异性免疫，从而增加免疫球蛋白的转移。然而，使用 FMT 进行活病原体暴露存在风险，需要通过进一步的研究来确定这种应用的最佳操作方法。后两种用途仅在动物医学领域得到应用，尚未在人体中开展试验。图 3-1-18 所示为粪便微生物群移植在家畜中的应用和预期结果，包括治疗、预防和免疫原性应用。

Therapeutic use of FMT is applied when the goal of administration is to treat clinical signs or resolve disease conditions that are ongoing. In contrast, FMT prophylaxis would be used to provide beneficial microbiome characteristics prior to high-risk of pathogen exposure. FMT has also been used as an immunostimulatory tool similar to vaccination, where the transplant material stimulates pathogen-specific immunity onset, as a part of preventative with the goal of increasing immunoglobulin transfer. However, risks exist in the use of FMT for live pathogen exposure, and additional research is needed to define the best practices for this application. Unique to veterinary medicine, the latter two uses have not been significantly explored in humans. Fig. 3-1-18 shows application and expected results of fecal microbiota transplantation in livestock, including therapeutic, preventive and immunogenic applications.

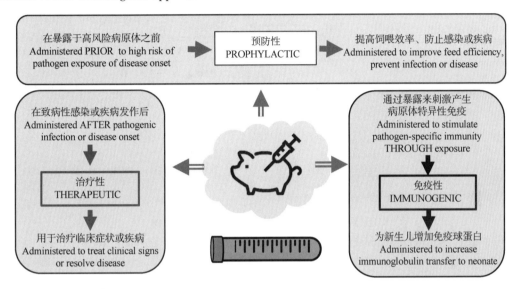

图 3-1-18 粪便微生物群移植在家畜中的应用和预期结果，包括治疗、预防和免疫原性应用
Fig. 3-1-18 Application and expected results of fecal microbiota transplantation in livestock, including therapeutic, preventive and immunogenic applications

Ⅱ 植物生物技术
Ⅱ Plant Biotechnology

2.1 增产
2.1 Yield Increase

2.1.1 提高固碳效率和作物产量
2.1.1 Improving Carbon Fixation Rate and Yield

提高固碳效率和产量的主要策略如图 3-2-1 所示。
Fig. 3-2-1 shows schematic summary of the main strategies for improving carbon fixation rate and yield.

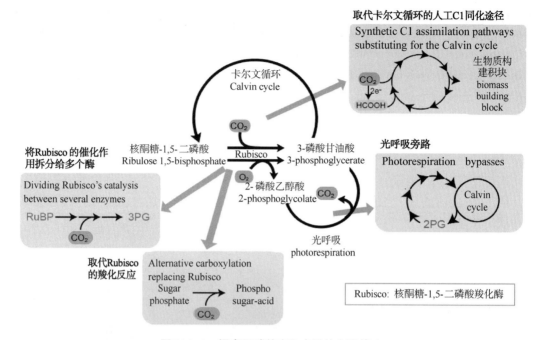

图 3-2-1　提高固碳效率和产量的主要策略

Fig. 3-2-1　Schematic summary of the main strategies for improving carbon fixation rate and yield

光合作用：绿色植物吸收阳光的能量，同化二氧化碳和水，制造有机物并释放氧气的过程。

Photosynthesis: the process that green plants absorb the energy of sunlight, assimilate carbon dioxide and water, produce organic matter and release oxygen.

光呼吸：植物的绿色细胞依赖光照，吸收 O_2 并放出 CO_2 的过程。

Photorespiration: the process that green cells in plants depend on light to absorb O_2 and release CO_2.

2.1.2　天然光呼吸及其合成旁路
2.1.2　Natural Photorespiration and Its Synthetic Bypass

图 3-2-2 所示为天然光呼吸途径和五个合成旁路。天然光呼吸途径用黑色箭头表示，其中棕色箭头突出了经济性极低的反应。甘油酸途径（紫色为叶绿体途径，粉红色为过氧化物酶体途径）和乙醇酸完全氧化途径（绿色箭头）已经在植物中进行了验证，但都导致了 CO_2 的释放。阿拉伯糖 5-磷酸途径（蓝色箭头）是一种碳保存途径，在不损失 CO_2 的情况下将乙醇酸同化到卡尔文循环（图3-2-3）。红色箭头显示了一个潜在的途径，利用叶绿体的低还原电位将 CO_2 还原为甲酸盐，甲酸盐随后被光呼吸途径吸收，从而将该途径转变为碳固定途径。

Fig. 3-2-2 shows schematic of native PR and five synthetic bypasses to this pathway. Native PR is shown by black arrows, where brown arrows highlight especially wasteful reactions. The glycerate pathway (chloroplastic version shown in purple and peroxisomal version shown in pink) and the glycolate complete oxidation pathways (green arrows) have already been tested in plants, but all result in the release of CO_2. The arabinose 5-phospate pathway (blue arrows) is a carbon-conserving route, assimilating glycolate to the Calvin cyde (Fig. 3-2-3) without the loss of CO_2. Red arrows show a potential pathway that harnesses the low reduction potential in the chloroplast to reduce CO_2 to formate, which is then assimilated into the photorespiratory pathway, transforming it into a carbon fixing route.

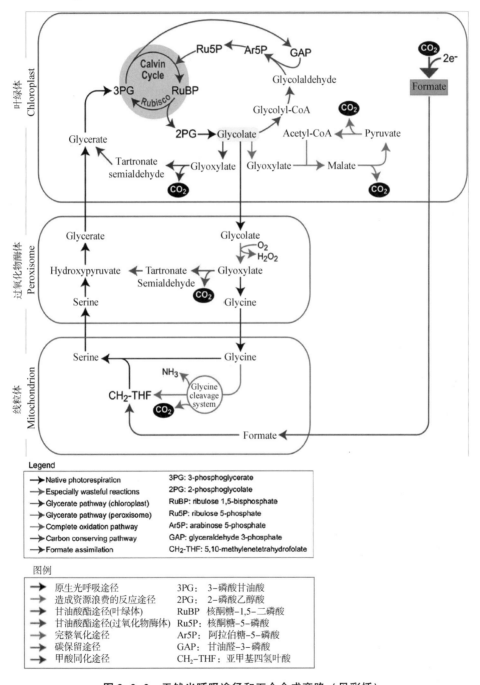

图 3-2-2 天然光呼吸途径和五个合成旁路（见彩插）

Fig. 3-2-2 Natural photorespiration and its synthetic bypasses (see the color figure)

2.1.3 卡尔文循环改造
2.1.3 Engineering of Calvin Cycle

图 3-2-3（a）所示为遮光处理时光保护与二氧化碳固定的相互作用。当叶片暴露于高光下时，二氧化碳固定速率较高，过量的激发能可通过非光化学猝灭（NPQ）无害地消散。NPQ 的水平与光系统 II 蛋白 PsbS 的丰度正相关，并在紫黄素脱环氧化酶（VDE）催化下的紫黄素向玉米黄质的转化中被进一步激活。在低光照处理下，二氧化碳固定会受到由光合作用电子传递产生的烟酰胺腺嘌呤二核苷酸磷酸和三磷酸腺苷的还原形式的限制，并转而受到高水平 NPQ 的限制。因此，在 NPQ 完全弛豫之前，二氧化碳固定速率一直处于受抑制状态。这一过程可持续数分钟到数小时，并与玉米黄质环氧化酶（ZEP）催化下的玉米黄素环氧化反应速率相关。图下方的文字描述了将 NPQ 弛豫加速至超过野生型（WT）烟草的策略。

Fig. 3-2-3 (a) shows interaction between photoprotection and CO_2 fixation during sun-shade transitions. When leaves are exposed to highlight, the rate of CO_2 fixation is high, and excessive excitation energy is harmlessly dissipated through non-photochemical quenching (NPQ). The level of NPQ is positively correlated with the abundance of photosystem II protein PsbS and further stimulated by the de-epoxidation of violaxanthin to zeaxanthin, catalyzed by violaxanthin de-epoxidase (VDE). Upon transition to lowlight, CO_2 fixation becomes limited by the reduced form of nicotinamide adenine dinucleotide phosphate and adenosine triphosphate derived from photosynthetic electron transport, which in turn is limited by high levels of NPQ. The rate of CO_2 fixation therefore remains depressed until relaxation of NPQ is complete. This can take minutes to hours and is correlated with the rate of zeaxanthin epoxidation, catalyzed by zeaxanthin epoxidase (ZEP). The text underneath the figure describes the strategy used to accelerate NPQ relaxation compared with WT tobacco.

图 2-3-3（b）所示为田间作物的产量。过表达 AtVDE、AtPsbS 和 AtZEP（VPZ）的品种比野生型植株产量高 15%。①总干质量。②叶面积。③植株高度。④试验田航空照片。

Fig. 2-3-3 (b) shows productivity of field-grown tabacum plants. Linesexpressing AtVDE, AtPsbS, and AtZEP (VPZ) produced 15% larger plants than the WT did. ①Total dry mass. ②Leaf area. ③Plant height. ④Aerial view of the field experiment.

如图 3-2-4 所示，CETCH 循环（CETCH 循环是由 17 种酶组成的反应网络，以每分钟每毫克蛋白质 5 纳摩尔的速度将二氧化碳转化为有机分子），包括校对酶和辅助因子再生酶。

As shown in Fig. 3-2-4, topology of the CETCH cycle (The CETCH cycle is a reaction network of 17 enzymes that converts CO_2 into organic molecules at a rate of 5 nanomoles of CO_2 per minute per milligram of protein), including proofreading and cofactor regenerating enzymes.

(a)
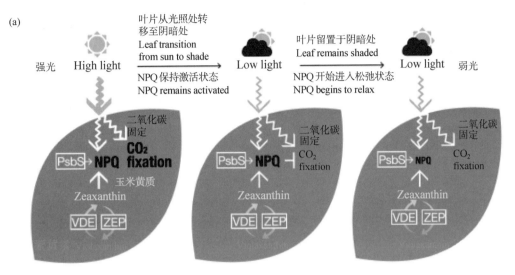

(b)
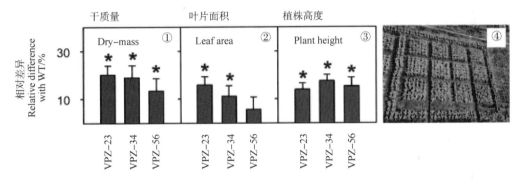

图 3-2-3　卡尔文循环改造

Fig. 3-2-3　Engineering of Calvin cycle

（a）遮光处理时光保护与二氧化碳固定的相互作用；（b）田间作物的产量

(a) Interaction between photoprotection and CO_2 fixation during sun-shade transitions;

(b) Productivity of field-grown tabacum plants

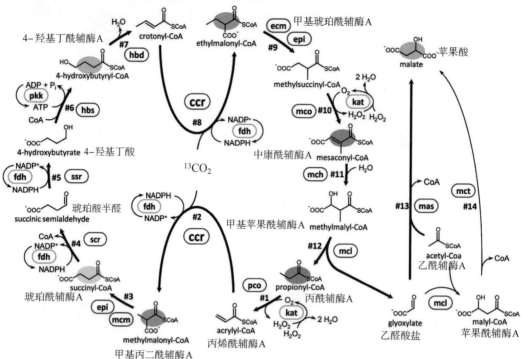

图 3-2-4　CETCH 循环

Fig. 3-2-4　CETCH cycle

2.2 抗虫作物
2.2 Insect Resistant Crops

转基因植物：利用基因工程技术改变基因组的组成以用于农业生产或农产品加工的植物。图 3-2-5 所示为转基因植物的构建及抗虫 Bt 基因的表达。

Genetically modified plants: plants that use genetic engineering technology to change the genome composition for agricultural production or agricultural product processing. Fig. 3-2-5 shows construction of transgenic plant and expression of Bt gene against insect larvae.

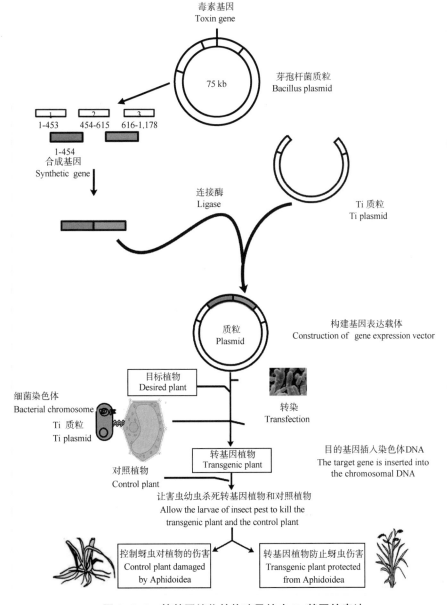

图 3-2-5 转基因植物的构建及抗虫 Bt 基因的表达

Fig. 3-2-5 Construction of transgenic plant and expression of Bt gene against insect larvae

2.3 作物品质提升
2.3 Crop Quality Improvement

图 3-2-6（a）所示为酵母中葫芦素 C 的合成途径。图 3-2-6（b）所示为提出的无苦味黄瓜的驯化模型。果实和叶子的苦味表型等位基因用不同的颜色表示。首先，Bt 基因调控区域的 SV-2195 和其他单核苷酸多态性变异使野生极苦黄瓜变成了条件性无苦味黄瓜，这类品种在受到环境胁迫后，果实仍然会变苦。下一步，携带 SNP-1601 变异的无苦味黄瓜品种被完全驯化；最后，葫芦素 C 合成途径的 Bi 变异又被用于现代黄瓜品种的选育。

Fig. 3-2-6 (a) shows biosynthetic pathway for cucurbitacin C production in S. cerevisiae. Fig. 3-2-6 (b) shows proposed model for the domestication of non-bitter cucumber. Various fruit and foliage bitterness phenotypes of gene alleles are indicated by different colors. First, SV-2195 and other SNP mutations within the regulatory region of Bt differentiated the extremely bitter ancient cucumber from other conditional non-bitter lines, whose fruit would turn bitter when grown under stress. Second, non-bitter cucumber carrying SNP-1601 were "fully" domesticated. Third, Bi, a mutation at the cucurbitacin C biosynthetic pathway is used in modern cucumber breeding.

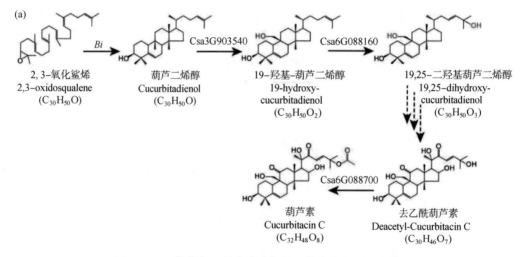

图 3-2-6 葫芦素 C 的合成途径及无苦味黄瓜的驯氏模型

Fig. 3-2-6 Biosynthetic pathway for cucurbitacin C production in S. cerevisiae and proposed model for the demestication of non-bitter cucumber

(a) 酵母中葫芦素 C 的合成途径

(a) Biosynthetic pathway for cucurbitacin C production in S. cerevisiae

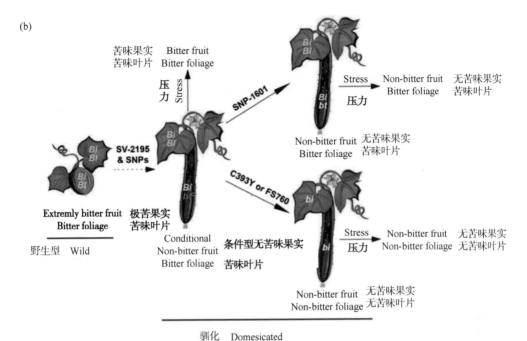

图 3-2-6 葫芦素 C 的合成途径及无苦味黄瓜的驯氏模型（续）
Fig. 3-2-6 Biosynthetic pathway for cucurbitacin C production in *S. cerevisiae* and proposed model for the demestication of non-bitter cucumber（continued）

(b) 提出的无苦味黄瓜的驯化模型
(b) Proposed model for the domestication of non-bitter cucumber

第4章 环境生物技术

Chapter 4　Environmental Biotechnology

本章提要

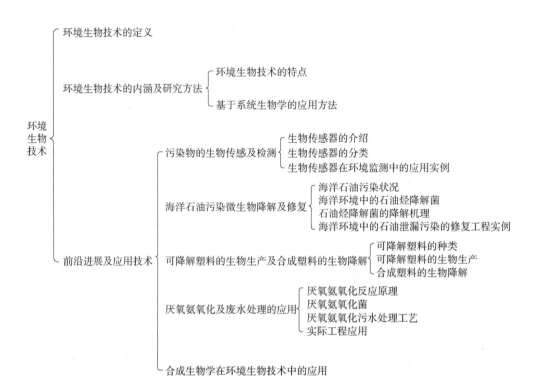

Summary of Chapter 4

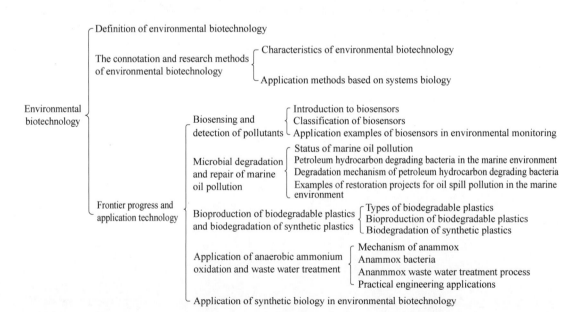

环境生物技术的定义
Definition of Environmental Biotechnology

环境生物技术（EBT）也可称为环境生物工程（environmental bioengineering），是近40年发展起来的一种现代生物技术与环境工程相结合的新兴交叉学科。广义上，环境生物技术是一切应用于环境污染治理方面的生物技术；目前，其已成为一种经济效益和环境效益俱佳、解决复杂环境污染问题的最有效的手段。狭义上，环境生物技术是直接或间接利用生物体的某些组成部分或某些机能建立降低或消除污染物产生的生产工艺，或者能够高效净化环境污染，同时生产有用物质的技术系统。

Environmental biotechnology (EBT), also known as environmental bioengineering, is an emerging interdisciplinary combination of modern biotechnology and environmental engineering that has developed over the past 40 years. In a broad sense, environmental biotechnology is all biotechnology applied to environmental pollution control; at present, it has become the most effective means to solve complex environmental pollution problems with good economic and environmental benefits. In the narrow sense, environmental biotechnology is a technical system that can directly or indirectly uses the living organisms or the active materials to establish a green process for producing useful substances with reduced or eliminated generation of pollutants or to transform the environmental pollutants to valuable materials.

2　环境生物技术的内涵及研究方法
2　The Connotation and Research Methods of Environmental Biotechnology

2.1　环境生物技术的特点
2.1　Characteristics of Environmental Biotechnology

根据环境生物技术的技术特点或精准程度将其分为三个层次，如图4-2-1所示。第一层次指利用天然（生态）处理系统进行污染治理的技术，如氧化塘、人工湿地系统等的特点是最大限度地发挥自然界的生物净化功能；第二层次指人工生物处理技术和工艺，如活性污泥法、生物膜法及其在新理论和技术背景下产生的强化生物处理技术和工艺等。第三层次指以基因工程为主导的现代污染防治生物技术，如基因工程菌构建、抗污染型转基因植物培育等。环境生物技术的这三个层次均是污染治理不可缺少的生物技术手段。第一层次的环境生物技术的最大特点是充分发挥自然界生物净化环境的功能，投资运行费用低，易于管理，是一种省力、省钱、省能的技术；第二层次的环境生物技术是当今废物生物处理中应用最广泛的技术，其技术本身也在不断改进，新的科学技术也在不断渗入，因此，它仍然是目前环境污染治理中的主力军；第三层次的环境生物技术需要以现代生物技术知识为背景，为寻求快速、有效的污染控制与预防新途径提供可能，是解决目前出现的日益严重且复杂的环境问题的强有力的手段。

According to the technical characteristics or accuracy of environmental biotechnology, it is divided into three levels, as shown in Fig. 4-2-1. The 1st level refers to technologies that use natural (ecological) treatment systems for pollution control, such as oxidation pond, constructed wetland system, etc., which are characterized by maximizing the biological purification function of nature. The 2nd level refers to artificial biological treatment technology and process, such as activated sludge method, biofilm method and the enhanced biological treatment technology produced under the background of new theories and technologies. The 3rd level refers to modern pollution prevention biotechnology led by genetic engineering, such as the construction of genetic engineering bacteria and the cultivation of anti-pollution transgenic plants. These three levels of environmental biotechnology are indispensable biotechnology tools for pollution control. The greatest features of the 1st level environmental biotechnology include the full function of using natural organisms to purify environment, the low investment operation cost, and the easy management, which lead environmental biotechnology to be a labor-saving, cost-saving and energy-saving technology. The 2nd level of environmental biotechnology is the most widely used technology in the biological treatment of waste. The technology itself is constantly improving, and new science and technology are also infiltrating. Therefore, it is still the main force in the current environmental pollution control. The 3rd level of environmental biotechnology needs to be based on modern biotechnology knowledge to provide a possibility for seeking new and effective ways of pollution control and prevention, and it is a powerful means to solve the increasingly serious and complex environmental problems.

在实际应用过程中,各种工艺与技术之间相互交叉、相互渗透,甚至将三个层次集为一体,在水污染控制、大气污染治理、有毒有害物质的降解、清洁可再生资源的开发、废物资源化、环境监测、污染环境的修复和工业企业的清洁生产等方面,发挥着极为重要的作用。

In the actual application process, the cross and penetration between various processes and technologies make the three levels integrated, which plays an extremely important role in the control of water and air pollution, the degradation of toxic and harmful substances, the development of clean renewable resources, the recycling of waste, environmental monitoring, the restoration of polluted environment and the clean production of industrial enterprises.

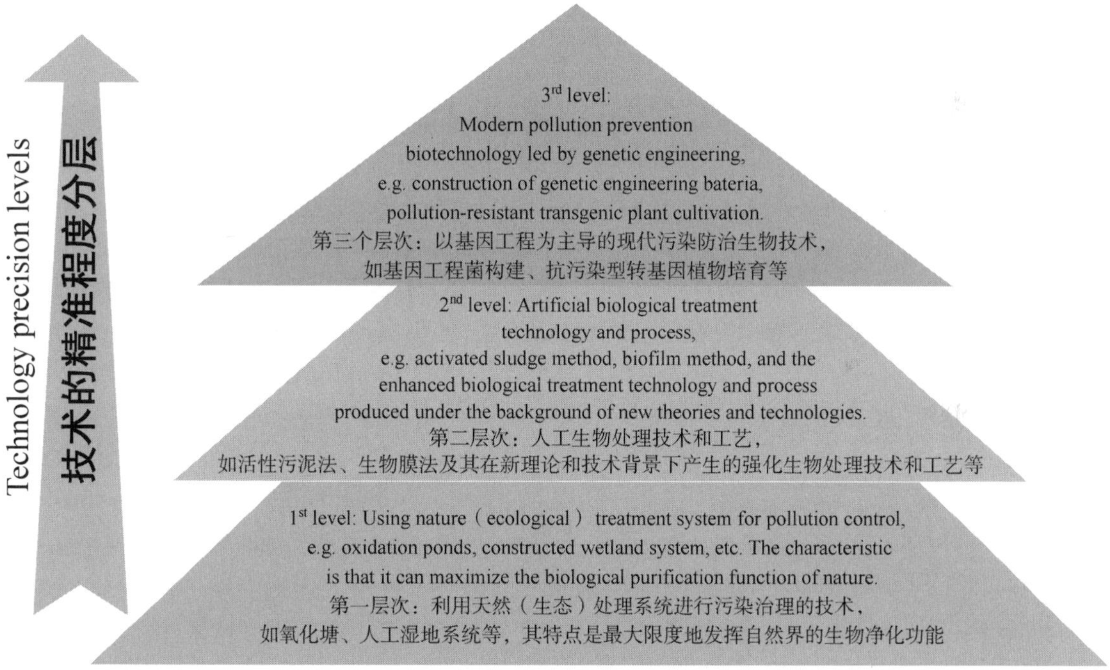

Technology precision levels / 技术的精准程度分层

3rd level:
Modern pollution prevention biotechnology led by genetic engineering,
e.g. construction of genetic engineering bateria,
pollution-resistant transgenic plant cultivation.
第三个层次:以基因工程为主导的现代污染防治生物技术,
如基因工程菌构建、抗污染型转基因植物培育等

2nd level: Artificial biological treatment technology and process,
e.g. activated sludge method, biofilm method, and the enhanced biological treatment technology and process produced under the background of new theories and technologies.
第二层次:人工生物处理技术和工艺,
如活性污泥法、生物膜法及其在新理论和技术背景下产生的强化生物处理技术和工艺等

1st level: Using nature (ecological) treatment system for pollution control,
e.g. oxidation ponds, constructed wetland system, etc. The characteristic is that it can maximize the biological purification function of nature.
第一层次:利用天然(生态)处理系统进行污染治理的技术,
如氧化塘、人工湿地系统等,其特点是最大限度地发挥自然界的生物净化功能

图 4-2-1　环境生物技术的三个层次

Fig. 4-2-1　Three levels of environmental biotechnology

2.2 基于系统生物学的应用方法
2.2 Application Methods Based on Systems Biology

生物修复是一种由微生物介导的过程，是一种降解和解毒环境污染物的可持续过程。系统生物学是一种综合研究方法，通过调查分子、细胞、群落和生态系统水平的相互作用和网络来研究复杂的生物系统。最近，基因组学、转录组学、蛋白质组学、代谢组学、表型组学和脂质组学的现代工具已被应用于无数环境中微生物群落的系统生物学的研究。

Bioremediation is a sustainable process mediated by microorganisms that degrades and detoxifies environmental pollutants. Systems biology is an integrated research approach to study complex biological systems, by investigating interactions and networks at the molecular, cellular, community, and ecosystem levels. Recently, modern tools of genomics, transcriptomics, proteomics, metabolomics, phenoomics, and lipidomics have been applied to myriad systems biology studies of environmental microbial communities.

基于微生物的环境生物技术对于修复污染环境尤为重要。要将系统生物学方法用于生物修复，就必须涉及微生物群落组成、细胞和分子活动的表征。生物修复的最终目标是消除或解毒有毒化合物，这需要了解环境变量和细胞与细胞间相互作用的所有可能影响。如果重点是阐明微生物群落组成，则应使用基于DNA的"组学"工具，如16S rRNA克隆文库，生物微矩阵芯片（PhyloChip）或测序。如果有兴趣了解细胞途径并鉴定参与微生物介导反应的功能基因，则应使用识别RNA的工具，以及诸如高通量基因芯片（GeoChip）、转录组测序技术（RNA-seq）和各种质谱方法的蛋白质。如果目的是表征微生物产生的小分子，可以使用基质辅助激光解吸/电离（MALDI），解吸电喷雾电离（DESI），核磁共振（NMR）光谱。对限制营养素、电子供体、电子受体和水文的伴随监测对于系统生物学概念模型也是至关重要的。

Microbial-based environmental biotechnology is particularly important for remediation of polluted environment. The use of systems biological approaches in bioremediation must involve characterization of microbial community composition, cellular and molecular activities. And achieving the ultimate goal of bioremediation, which is to eliminate or detoxify toxic compounds, requires understanding all possible effects of environmental variables and cell-cell interactions. If the focus is on elucidating microbial community composition, DNA based "omics" tools such as 16S rRNA clone library, PhyloChip or sequencing should be used. If the interest is to understand cellular pathways and identify functional genes involved in microbial-mediated responses, tools for RNA recognition and proteins such as GeoChip, RNA-seq, and various mass spectrometry methods should be used. If the intention is to characterize small molecules produced by the microbes, matrix-assisted laser desorption/ionization (MALDI), desorption electrospray ionization (DESI), nuclear magnetic resonance (NMR) spectroscopy can be used. Concomitant monitoring of limiting nutrients, electron donors, electron acceptors, and hydrology is also crucial for a systems biology conceptual model.

一般而言，生态系统由群落、种群、细胞、蛋白质、脂质和代谢物，DNA和RNA组成，如图4-2-2。这些方法使用地球化学、生态学、基因组学、蛋白质组学、代谢组学和计算技术，在细胞水平上分析 DNA、RNA 和蛋白质，以了解生物修复功能如何影响细胞，分析群落和群体，了解生物修复实践对生态系统层面的结构/功能关系和最终的相互作用。

In general, an ecosystem is composed of communities, populations, cells, protein, lipid and metabolite, RNA, and DNA, as shown in Fig. 4-2-2. The approaches use geochemical, ecological, genomic, proteomic, metabolomic, and computational techniques to analyze DNA, RNA, and protein at the cellular levels to understand how bioremediation functions affect cells, to analyze communities and populations, and to understand structure/function relationships and final interactions between bioremediation practice and the ecosystem.

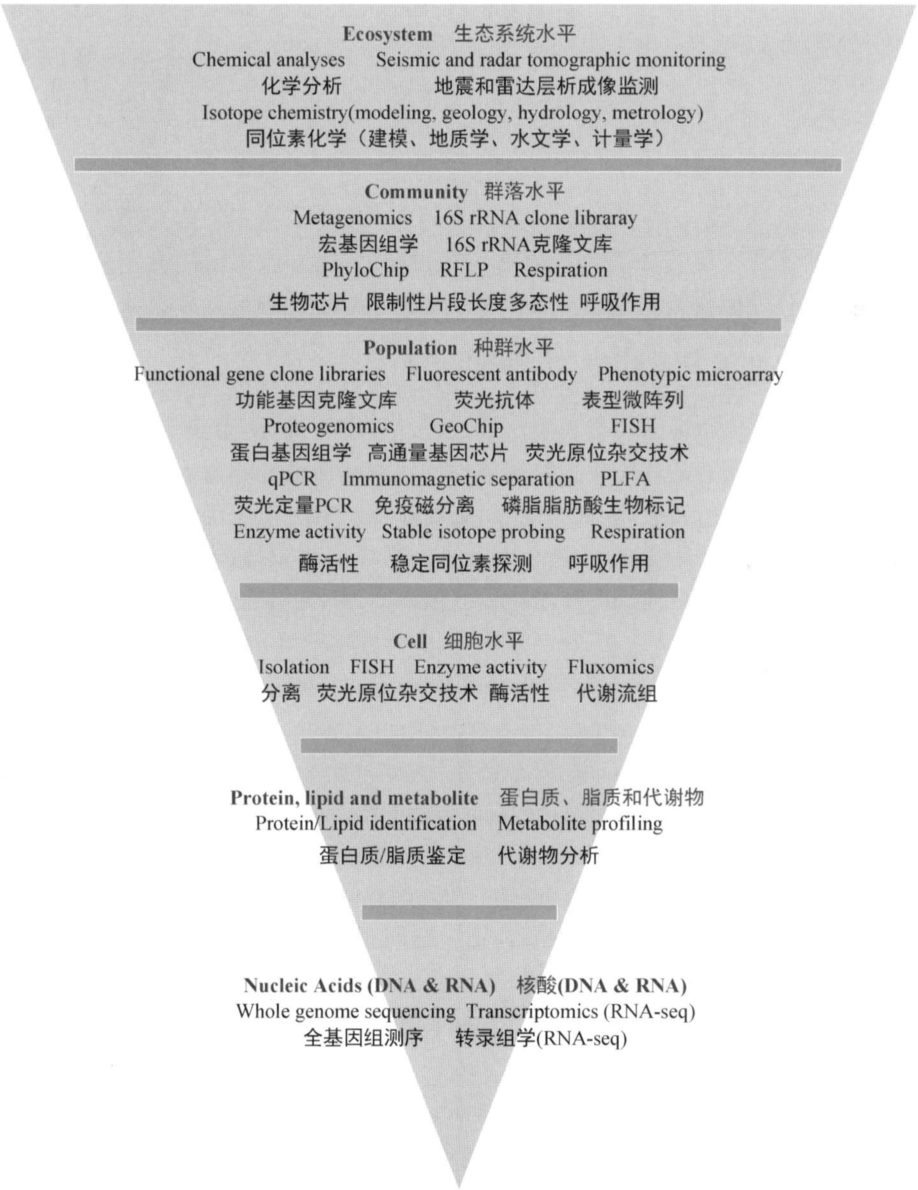

Ecosystem 生态系统水平
Chemical analyses Seismic and radar tomographic monitoring
化学分析 地震和雷达层析成像监测
Isotope chemistry(modeling, geology, hydrology, metrology)
同位素化学（建模、地质学、水文学、计量学）

Community 群落水平
Metagenomics 16S rRNA clone libraray
宏基因组学 16S rRNA克隆文库
PhyloChip RFLP Respiration
生物芯片 限制性片段长度多态性 呼吸作用

Population 种群水平
Functional gene clone libraries Fluorescent antibody Phenotypic microarray
功能基因克隆文库 荧光抗体 表型微阵列
Proteogenomics GeoChip FISH
蛋白基因组学 高通量基因芯片 荧光原位杂交技术
qPCR Immunomagnetic separation PLFA
荧光定量PCR 免疫磁分离 磷脂脂肪酸生物标记
Enzyme activity Stable isotope probing Respiration
酶活性 稳定同位素探测 呼吸作用

Cell 细胞水平
Isolation FISH Enzyme activity Fluxomics
分离 荧光原位杂交技术 酶活性 代谢流组

Protein, lipid and metabolite 蛋白质、脂质和代谢物
Protein/Lipid identification Metabolite profiling
蛋白质/脂质鉴定 代谢物分析

Nucleic Acids (DNA & RNA) 核酸(DNA & RNA)
Whole genome sequencing Transcriptomics (RNA-seq)
全基因组测序 转录组学(RNA-seq)

图 4-2-2　从分子到生态系统的系统生物学
Fig. 4-2-2　System biology from molecule to ecosystem

Ⅲ 前沿进展及应用技术
Ⅲ Frontier Progress and Application Technology

近年来，环境生物技术发展极其迅猛。无论是在生态环境保护方面，还是在污染预防和治理方面以及环境监测方面，都显示出独特的功能和优越性。环境生物技术在科学与工程中的应用极其广泛，包括环境污染防治、可再生能源开发、废物资源化、清洁生产等方面。此外，生物科学和生物技术的发展将为环境生物技术开辟更加广阔的发展前景。

In recent years, the development of environmental biotechnology is extremely rapid, and it has shown unique functions and advantages in ecological environment protection, pollution prevention and treatment, and environmental monitoring. Environmental biotechnology is widely used in science and engineering, including environmental pollution prevention, renewable energy development, waste recycling, clean production, etc. Besides, the development of biological science and biotechnology will open up a broader prospect for environmental biotechnology.

3.1 污染物的生物传感及检测
3.1 Biosensing and Detection of Pollutants

环境生物技术不仅单纯适用于环境污染治理，如今已相当广泛地应用于环境监测，尤其是以生物传感器为核心的环境生物监测技术。生物传感器可以满足实施自动和连续监测的需要，判断环境污染的发展趋势，探索污染物在环境中的迁移转化以及降解规律，监测污染致突变的成因，分析污染的来源，从而使生物环境污染监测更便捷、更灵敏、更全面。生物传感器可在线迅速地提供环境质量参数，成为环境质量预报和报警中的重要组成部分。

Environmental biotechnology is not only suitable for environmental pollution control, Environmental biotechnology is not only suitable for environmental pollution control, but also widely used in environmental monitoring, especially biosensor based environmental biological monitoring technology. Biosensors can meet the needs of automatic and continuous monitoring, judge the development trend of environmental pollution, explore the migration and transformation of pollutants in the environment and its degradation rules, monitor the cause of pollution mutagenesis, analyze the source of pollution, so as to make biological environmental pollution monitoring more convenient, more sensitive and more comprehensive. Biosensors can quickly provide environmental quality parameters online, which makes it become an important part of environmental quality predictions and alarms.

传统的生物环境监测主要有以下几种方法：用细菌总数及粪便污染指示菌监测水质；用鼠伤寒沙门菌检验物质致突变性与致癌性；用发光细菌快速检测环境毒物；通过水中藻类的生长量监测水质或检测物质的毒性。这些方法在操作及检验标准上均已成熟，已普遍使用。近年来，随着技术的进步和理论的发展，一些新的监测方法不断涌现出来。例如，利用核酸杂交技术检测水环境中的致病菌（大肠杆菌、志贺菌、沙门菌、耶尔森菌等腹泻性致病菌）；利用 PCR 技术监测土壤、沉积物、水样等环境中尚不能培养的微生物；利用酶联免疫分析法监测环境中的杀虫剂、杀菌剂、除草剂等农药以及多氯联苯、二噁英、抗菌素等污染物；用于水质监测的 BOD 传感器、硝酸盐微生物传感器、酚类以及阴离子表面活性剂传感器和水体富营养化监测传感器；用于大气和废气监测的亚硫酸、亚硝酸盐、氨、甲烷及一氧化碳、微生物传感器等。

Traditional biological environmental monitoring mainly includes the following methods: monitoring the water quality with the total number of bacteria and fecal contamination indicator bacteria, detecting mutagenicity and carcinogenicity with *Salmonella typhimurium*, rapidly detecting environmental poisons with luminescent bacteria, monitoring water quality or detecting the toxicity with the growth of algae in water. These methods are mature and widely used in operation and inspection standards. In recent years, technological progress and theoretical development have brought about some new monitoring methods. For example, nucleic acid hybridization technology is used to detect pathogenic bacteria (diarrheal pathogenic bacteria such as *Escherichia coli*, Shigella, Salmonella, Yersinia) in water, PCR is used to monitor microorganisms that cannot be cultivated in the environment such as soil, sediment, water samples, etc., enzyme-linked immunosorbent assay is used to monitor pesticides, fungicides, herbicides and pollutants such as PCBs, dioxins, and antibiotics in the environment. BOD sensors, nitrate microbial sensors, phenolic and anionic surfactant sensors and water eutrophication monitoring sensors are used to monitor water quality, and sulfurous acid, nitrite, ammonia, methane and carbon monoxide, microbial sensors, etc., for atmospheric and exhaust gas monitoring.

3.1.1 生物传感器的介绍
3.1.1 Introduction to Biosensors

生物传感器由生物来源的识别元件（如酶、抗原、抗体、细胞、微生物、组织、核酸等）和"第二重要"的换能器紧密连接或集成。目标分析物与固定在支持基质上的互补生物识别元素特异性相互作用导致物理化学性质的变化，包括传热、pH 变化、质量变化、电子转移以及气体或特定离子的吸收或释放。通过换能器以电子信号的形式检测和测量这些变化，该电子信号与结合的分析物的浓度成比例。生物传感器技术就是利用产生信号的形式测量生物输出并监测。图 4-3-1 所示为生物传感器的发展情况。

Biosensors are tightly coupled or integrated with biologically derived recognition elements (e.g., enzymes, antigens, antibodies, cells, microorganisms, tissues, nucleic acids, etc.) and "the 2nd most important" transducers. Specific interaction of the target analyte with complementary biorecognition elements immobilized on the support matrix results in changes in physicochemical properties, including heat transfer, pH changes, mass changes, electron transfer, and absorption or release of gases or specific ions. The electronic signal generated by the transducer is proportional to the concentration of the bound analyte, and these changes can be detected and measured, so biosensor technology uses signal generation to detect and measure biological output. Fig. 4-3-1 shows the development of bosensors.

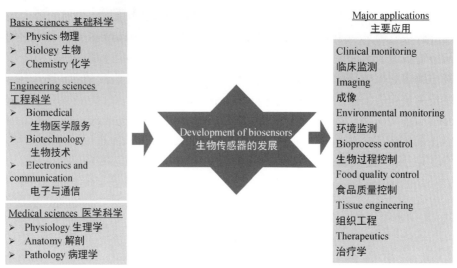

图 4-3-1　生物传感器的发展

Fig. 4-3-1　The development of biosensors

如图 4-3-2 所示，生物传感器系统中的组成及其功能包括以下五个。

（1）识别元件——生物受体特异性地与分析物结合。

（2）电子接口——在该接口处发生特定的生物过程，从而产生信号。

（3）传感器元件——特定的生化反应被转换为电子信号。

（4）信号处理元件——电子信号转换为物理参数。

（5）显示单元——向操作员显示结果的界面。

As shown in Fig. 4-3-2, the five different components and their functions in a biosensor system include the following five parts:

（1）Recognition element—bioreceptors bind specifically to the analyte.

（2）Electrical interface—specific biological processes occur at this interface which give rise to a signal.

（3）Transducer element—specific biochemical reaction is being converted into electrical signal.

（4）Signal processor—electronic signal is converted into a physical parameter.

（5）Display unit—an interface to display the results to the operator.

随着基因工程技术的出现，可以使用分子工具通过以下方式改进并构建更灵活、更有效的生物传感器。

（1）构建生产表面活性剂的微生物，以提高污染物的生物利用率。

（2）在降解污染物中构建更有效的微生物。

（3）构建能够利用多种化合物的微生物。

（4）构建能够经受恶劣环境条件的微生物，如极高温度和极低温度、高盐度或氧气含量有限的地区。

（5）选择具有新酶功能的微生物。

With the advent of genetic engineering technologies, more flexible and efficient biosensors can be improved by using molecular tools in the following ways.

(1) Construct microorganisms that produce surfactants to increase the bioavailability of pollutants.

(2) Construct more efficient microorganisms in degrading pollutant.

(3) Construct microorganisms which are capable of utilizing multiple types of compounds.

(4) Construct microorganisms which can survive in harsh environment conditions like extremely high and low temperature, high salinity, or areas with limited oxygen content.

(5) Select microorganisms with new enzymatic capabilities.

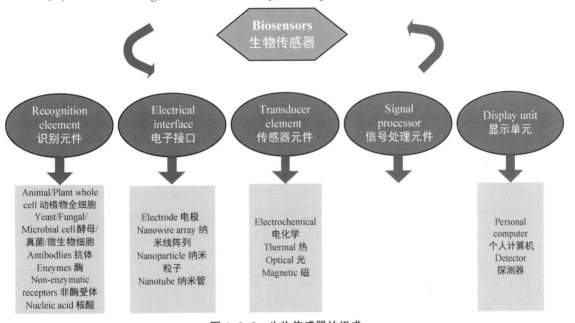

图 4-3-2　生物传感器的组成

Fig. 4-3-2　Components of the biosensor

3.1.2　生物传感器的分类
3.1.2　Classification of Biosensors

生物传感器一般可以从以下三个方面进行分类。

(1) 根据传感器输出信号的产生方式，可分为生物亲和型生物传感器和催化型生物传感器。

(2) 根据生物传感器的信号转化器可分为电化学生物传感器、半导体生物传感器、测热型生物传感器、测光型生物传感器、测声型生物传感器等。

(3) 根据生物传感器中分子识别元件上的敏感物质可分为酶传感器、微生物传感器、细胞传感器、动植物组织传感器、免疫传感器等。

Biosensors can generally be classified in the following three ways.

(1) According to the generation mode of sensor output signal, it can be divided into bio-affinity biosensor and catalytic biosensor.

(2) According to the biosensor signal generating method, it can be divided into electrochemical biosensor, semiconductor biosensor, calorimetric biosensor, photometric biosensor, and sound biosensor.

(3) According to the sensitive substance on the molecular recognition component in the biosensor, It can be divided into enzyme sensor, microbial sensor, cell sensor, animal and plant tissue sensor, immunosensor, etc.

A. 酶传感器

A. Enzyme sensors

酶传感器是利用酶的催化作用，在常温常压下将糖类、醇类、有机酸、氨基酸等生物分子氧化或分解，然后通过换能器将反应过程中化学物质转变的信号记录下来，进而推出相应的生物分子浓度。酶传感器的工作原理（图4-3-3）：在电化学电极的顶端紧贴一层酶膜就成了一支酶电极，溶液中的待测物质通过扩散进入酶膜，发生酶促反应，产生或消耗一种电活性物质，这种物质与待测物之间具有严格的化学计量关系。电活性物质的产生或消耗由电极监测。依据输出信号的不同，电极可分为电流型和电位型两种，前者输出电流，后者输出电压，电流和电压的变化与电活性物质的浓度相关。

The enzyme sensor utilizes enzyme catalysis to oxidize or decompose biomolecules such as sugars, alcohols, organic acids and amino acids under normal temperature and pressure, and then records the signal of chemical substances transformation in the reaction process through a transducer, which can help to get the corresponding biomolecule concentration. The working principle of enzyme sensors (Fig. 4-3-3) is that an enzyme electrode is made by sticking a layer of enzyme membrane on the top of the electrochemical electrode. The substance to be tested in the solution enters the enzyme membrane by diffusion and an enzymatic reaction occurs to generate or consume an electroactive substance. Because of the strict stoichiometric relationship between the electroactive substance and the substance to be measured, the electrode can monitor the production or consumption of the electroactive substance. According to the different types of output signals, electrodes can be divided into two types, current type of output current and potential type of output voltage, and the concentration of electroactive substances is related to the change of current and voltage.

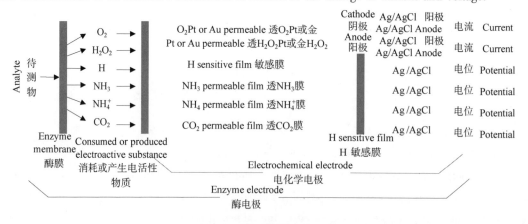

图 4-3-3 酶传感器的工作原理

图 4-3-3 Working principle of enzyme sensors

B. 微生物传感器

B. Microbial sensors

微生物传感器是由固定化微生物膜和换能器紧密结合而成的。常用的微生物有细菌和酵母菌。微生物的固定方法主要有吸附法、包埋法、共价交联法等。载体有胶原、醋酸纤维素和聚丙烯酰胺凝胶等。微生物传感器以活的微生物作为敏感材料,利用其体内的各种酶系及代谢系统来测定和识别相应底物。图4-3-4 所示为微生物传感器的工作原理。

The microbial sensors are a tightly integrated combination of an immobilized microbial membrane and a transducer. Commonly used microorganisms are bacteria and yeast. The immobilization methods of microorganisms mainly include adsorption method, embedding method, covalent cross-linking method, etc. The carriers are collagen, cellulose acetate, polyacrylamide gel, etc. Microbial sensors use living microorganisms as sensitive materials, and use various enzyme systems and metabolic systems in their bodies to determine and identify corresponding substrates. Fig. 4-3-4 shows the working principle of mocrobial sensors.

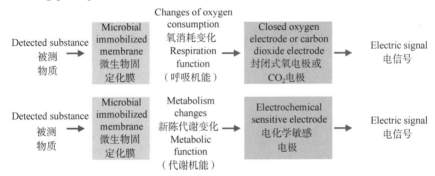

图 4-3-4 微生物传感器的工作原理

Fig. 4-3-4 Working principle of microbial sensors

根据微生物与底物作用原理,微生物电极可分为两种:一种是测定呼吸活性型微生物电极——微生物与底物作用,在同化样品中有机物的同时,微生物细胞的呼吸活性有所提高,依据反应中氧的消耗或二氧化碳的生成来检测被微生物同化的有机物的浓度;另一种是测定代谢物质型微生物电极——微生物与底物作用后生成各种电极敏感代谢产物,利用对某种代谢产物敏感的电极即可检测原底物的浓度。

According to the principle of the interaction between microorganisms and substrates, microbial electrodes can be divided into two types. One is microbial electrodes for the determination of respiratory activity—microorganisms react with substrates to improve the respiratory activity of microbial cells while assimilating organic matter in samples. The concentration of organic substance assimilated by microorganisms is measured by oxygen consumption or carbon dioxide production in the reaction. The other one is microbial electrodes for determination of metabolites—various electrode sensitive metabolites are generated after the interaction between microorganisms and substrates, and the concentration of the original substrate can be detected by using electrodes sensitive to certain metabolites.

C. 细胞传感器

C. Cell sensors

细胞传感器（图4-3-5）由基于重组质粒或全细胞的生物组分组成。它包含报告基因，对靶分子敏感的（分析物）传感元件基因，以及在靶分子存在或不存在下可以打开或关闭的启动子。要求报告系统产生信号，其强度与启动子的表达成正比。从通用操纵子中选择并与报告系统融合的启动子可以基于生物组分与靶分子的特异性相互作用打开或关闭并产生信号。

Cell sensors (Fig. 4-3-5) consist of biological components based on recombinant plasmids or whole cells. It has reporter gene, sensing element gene that are sensitive to target molecules (analytes), and promoters that can be turned on or off in the presence or absence of target molecules. The reporter system is required to produce a signal whose strength is proportional to the expression of the promoter. Promoter selected from a generic operon and fused with a reporter system, can be made turn on or off based on the specific interaction of the biological component with the target molecule and generate signal.

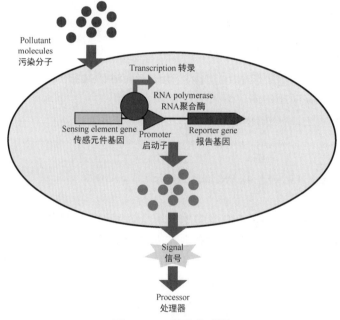

图 4-3-5 细胞传感器

Fig. 4-3-5 The cell sensor

特定基因的报告系统代码是表达载体的一部分，催化生化反应以产生信号。一些常用的报告系统包括细菌荧光素酶和绿色荧光蛋白（GFP）报告系统。细菌荧光素酶报告系统的活性取决于生物发光形式的光发射，生物发光是荧光素酶（由 *lux* 操纵子编码并与启动子和传感元件基因融合）活性的酶促反应。可以通过光电倍增管接收发射的光用于信号分析。GFP 具有内部发色团，赋予其荧光特性，因此在用紫外线或蓝光激发时发出明亮的绿光。

The reporter system code for a specific gene is part of an expression vector that catalyzes biochemical reactions to produce a signal. Commonly used reporter systems are bacterial luciferase and green fluorescent protein (GFP) reporter system, whose activity depends on light emission in the form of bioluminescence. Bioluminescence is an enzymatic reaction catalyzed by luciferase (encoded by the *lux* operon and fused with initiator and sensing element genes), that can receive emitted light through photomultiplier tubes for signal analysis. GFP has an internal chromophore, which gives it fluorescence properties and thus gives off a bright green light when excited with ultraviolet or blue light.

3.1.3 生物传感器在环境监测中的应用实例
3.1.3 Application Examples of Biosensors in Environmental Monitoring

生物传感器是一种很有前途的分析工具，使用简单、特异、经济、可靠，并提供可重复的结果。生物传感器的性能取决于生物元件的性质和稳定性、生物分子的固定方法、分析物的特定类型、所用换能器的类型、分析物的理化性质和操作条件。关键基质的检测无须事先分离、高选择性、ng/mL 或 pg/mL 的灵敏度、响应时间短、数据收集速度快、成本效益低这些区别于传统分析的优势使生物传感器在生物医学和环境监测中使用广泛。

Biosensors are a promising analytical tool that is simple to use, specific, cost-effective, reliable, and provides reproducible results. The biosensor performance depends upon the nature and stability of biological elements, method for immobilization of biomolecules, specific type of analyte, type of transducer used, physiochemical properties of analyte, and operating conditions. Biosensor is widely used in biomedical and environmental monitoring because of its advantages over traditional assays, such as the detection of critical substrates without prior isolation, high selectivity, sensitivity of ng/mL or pg/mL, short response time, fast data collection, and low-cost effectiveness.

适体传感器前景广泛，它利用适体作为识别元素。适体的优异特性和领先的检测平台技术（如光学、电化学与纳米材料集成，或质量敏感技术）相结合，具有高灵敏度和特异性，为适用于检测有害的小型毒性化学品和在环境中进行实时监测的适体传感器应用提供了希望。

Aptasensors have a broad prospect with aptamers as its recognition elements. The aptasensor can be highly sensitive and specific by combining excellent property aptamers with leading detection platform technologies such as optical, electrochemical and nanomaterial integration or mass sensitive technologies, which provides hope for the development of aptasnesors suitable for the detection of harmful small toxic chemicals and for real-time monitoring in environment.

图 4-3-6 所示为光学适体传感器检测策略说明。

Fig. 4-3-6 shows illustration of detection strategies of optical aptasensors.

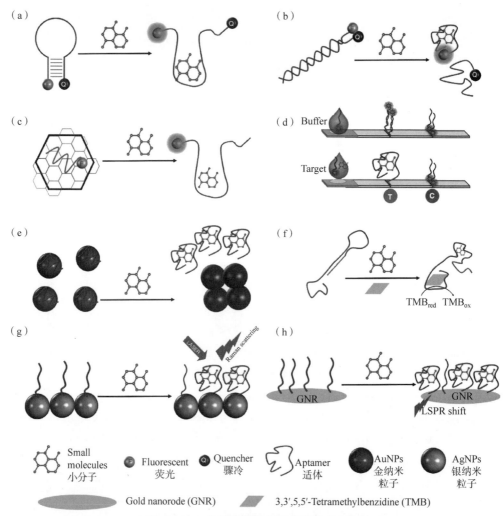

图 4-3-6 光学适体传感器检测策略说明

Fig. 4-3-6 Illustration of detection strategies of optical aptasensors

(a) 基于分子信标的适体传感器；(b) 基于互补序列的适体传感器；
(c) 基于石墨烯/CNT 的适体传感器；(d) 侧向流动测定（LFA）适体传感器；
(e) 金纳米粒子（AuNP）基于聚集的适体传感器；(f) 基于 DNA 酶的适体传感器；
(g) 基于表面增强拉曼散射（SERS）的适体传感器；
(h) 基于局部表面等离子共振（LSPR）移位的适体传感器

(a) Molecular beacon-based aptasensor; (b) Complementary sequence-based aptasensor;
(c) Graphene/CNT-based aptasensor; (d) Lateral flow assay (LFA) aptasensor;
(e) Gold-nano-particle (AuNP) aggregation-based aptasensor; (f) DNAzyme-based aptasensor;
(g) Surface-enhanced Raman scattering (SERS)-based aptasensor;
(h) Localized surface plasmon resonance (LSPR) shift-based aptasensor

图 4-3-7 所示为不同电化学适体传感器的说明。
Fig. 4-3-7 shows illustration of different electrochemical aptasensors.

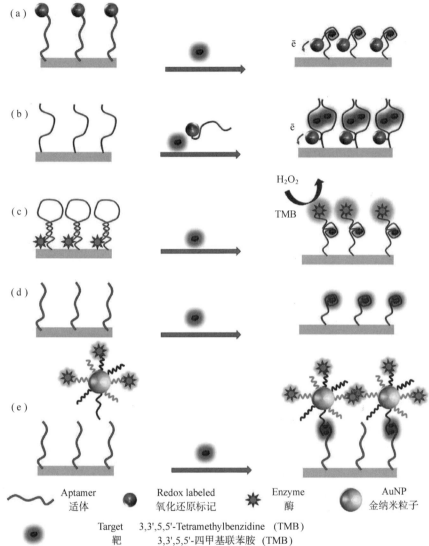

图 4-3-7　不同电化学适体传感器的说明

Fig. 4-3-7　Illustration of different electrochemical aptasensors

（a）标记的适体基电化学适体传感器；（b）标记/未标记的基于分离适体的电化学适体传感器；
（c）基于酶的电化学适体传感器；（d）基于标记的适体电化学适体传感器；
（e）基于纳米材料的电化学适体传感器

(a) Labeled aptamer-based electrochemical aptasensors;
(b) Labeled/Unlabeled split aptamer-based electrochemical aptasensors;
(c) Enzyme-based electrochemical aptasensors;
(d) Label free aptamer-based electrochemical aptasensors;
(e) Nanomaterial-based electrochemical aptasensors

图 4-3-8 所示为基于纳米材料的电化学适体传感器。
Fig. 4-3-8 illustrates nanomaterial-based electrochemical aptasensors.

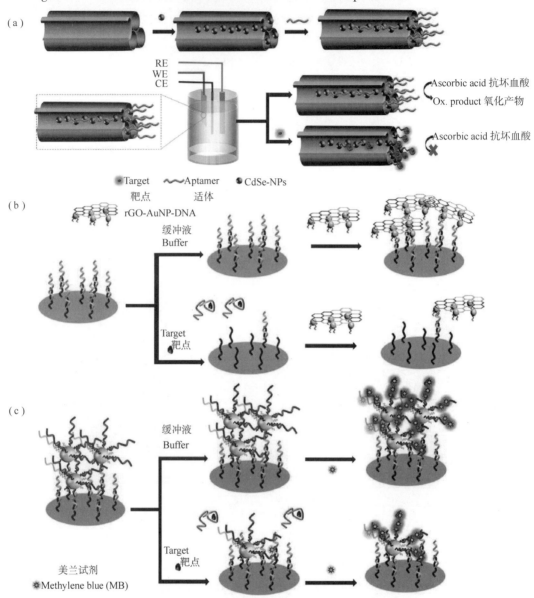

图 4-3-8 基于纳米材料的电化学适体传感器
Fig. 4-3-8 Nanomaterial-based electrochemical aptasensors

(a) 光电化学 (PEC) CdSe NPs 传感平台（工作电极 (WE)，参比电极 (RE) 和反电极 (CE)）；
(b) 基于 rGO 纳米片的适体传感器；(c) AuNP 电化学适体传感器

Fig. 4-3-8 (a) Photoelectrochemical (PEC) CdSe NPs sensing platform (working electrode (WE), reference electrode (RE) and counter electrode (CE));
(b) rGO nanosheet-based aptasensor; (c) AuNP electrochemical aptasensor

水中已知或未知的污染物在细胞中引起各种毒性作用，这可以表现为细胞分离、形态变化、细胞死亡，并且在色素细胞的情况下，表现为色素转移。这些细胞毒性信号可以在生物传感器中通过如细胞阻抗或色素追踪传导到整合的接收端以指示毒性。细胞生物传感器能够检测广谱物质，并可应用于水中毒性检测的许多不同领域，且这些应用可能有一些共同目标。图 4-3-9 所示为用于检测水中毒性的细胞生物传感器。

Known or unknown contaminants in water elicit various toxic effects in cells, which can be manifested as cell detachment, morphology change, cell death, and in chromatophores, it manifests as pigment transfer. These cytotoxic signals can be transduced to the integrated receiving end in biosensors by, for example cell impedance or pigment tracking, to indicate toxicity. Cell biosensors are able to detect a broad spectrum of substances and can be applied in many different fields of toxicity detection in water. Some of the applications may share some common aims (overlapping circles). Fig. 4-3-9 shows cell biosensors for detecting toxicity in water.

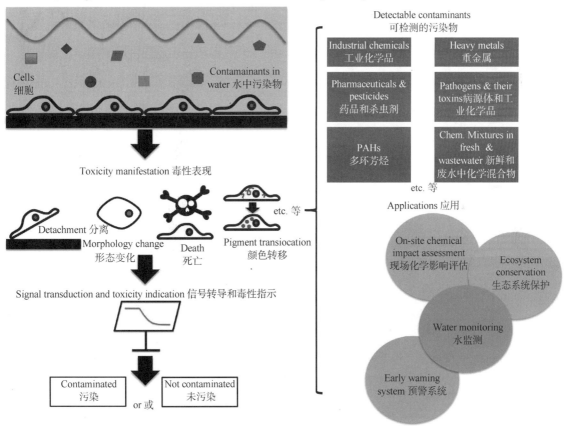

图 4-3-9　用于检测水中毒性的细胞生物传感器
Fig. 4-3-9　Cell biosensors for detecting toxicity in water

金属离子的现场和实时检测对于环境监测和了解金属离子对人类健康的影响非常重要。然而，开发对可以在未处理的样品和细胞的复杂基质中起作用的各种金属离子具有选择性的传感器提出了重大挑战。为了应对这些挑战，DNA 酶是一种新兴的金属离子依赖性酶，几乎可用于任何金属离子，已经用荧光团、纳米粒子和其他成像剂进行功能化，并被纳入传感器中，用于检测环境样品中的金属离子和成像活细胞中的金属离子。如图 4-3-10（a）所示，DNA 酶是通过体外选择的迭代组合选择策略选择的。

Real-time field detection of metal ions is very important for environmental monitoring and understanding the effects of metal ions on human health. However, there are significant challenges in developing sensors that are selective for various metal ions in untreated samples or the complex matrix of cells. DNAse, an emerging metal ion-dependent enzyme that can adapt to almost any metal ion, could address these challenges. It has been functionalized with fluorophores, nanoparticles, and other imaging agents and incorporated into sensors for the detection of metal ions in environmental samples and in imaging living cells. DNAses are selected via an iterative combinatorial selection strategy called *in vitro* selection as shown in Fig. 4-3-10 (a).

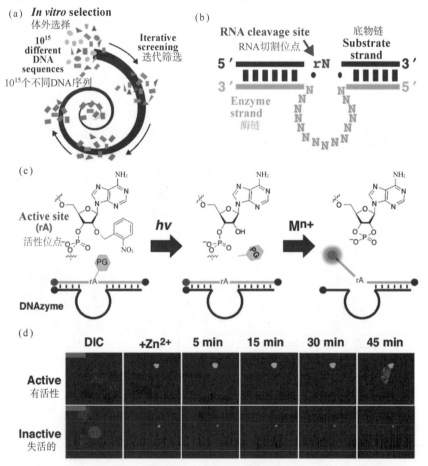

图 4-3-10 通过体外选择的迭代组合选择策略选择 DNA 酶（见彩插）

Fig. 4-3-10 DNAses are selected via an iterative combinatorial selection strategy called *in vitro* selection (see the color figure)

(a) DNA 酶通过体外选择的迭代组合选择策略选择；(b) 具有根据 Watson-Crick 碱基配对杂交的结合臂 DNA 酶；(c) DNA 酶被转变成一个催化信标；(d) 光敏在代信标的有效示例

(a) DNAses are selected via an iterative combinatorial selection strategy called *in vitro* selection;
(b) The resulting DNase has binding arms that hybridize according to Watson-Crick base pairing;
(c) DNAse turned into a catalytic beacon; (d) An example of the efficacy of the photocaged catalytic beacon

如图 4-3-10 （b） 所示，得到的 DNA 酶具有根据 Watson-Crick 碱基配对杂交的结合臂，由黑色和绿色表示，其是高度可编程的一个酶区域，显示为重复的 N（其中 N 是 A、C、G 或 T）和 rN 处的红色 RNA 切割位点（其中 N 可以是 A，C，G 或 U）。如图 4-3-10 （c） 所示，通过用荧光团（红色）和猝灭剂（灰色）功能化的 DNA 酶被转变成一个催化信标，在选择性单价、二价或三价金属离子辅助因子存在下去开启荧光信号。催化信标可以使用光敏的硝基苄基基团在 2′-OH 进行封闭。图 4-3-10 （d） 展示了光敏在代信标的有效示例。

Fig. 4-3-10 shows the resulting DNase has binding arms that hybridize according to Watson-Crick base pairing, indicated by black and green bars. This binding arm is a highly programmable enzymatic region shown as a red RNA cleavage site at repeated N (N is A, C, G, or T) and rN (N can be A, C, G, or U). As shown in Fig. 4-3-10 （c）, By functionalizing the DNAse with fluorophores (red) and quenchers (grey), it can be turned into a catalytic beacon, which can turn on fluorescent signal in the presence of selective monovalent, divalent, or trivalent metal-ion cofactor. The catalytic beacon can be caged by using the photolabile nitrobenzyl group on the 2′-OH. An example of the efficacy of the photocaged catalytic beacon is shown in Fig. 4-3-10 （d）.

3.2 海洋石油污染微生物降解及修复
3.2 Microbial Degradation and Repair of Marine Oil Pollution

3.2.1 海洋石油污染状况
3.2.1 Status of Marine Oil Pollution

海洋占了地球表面积的 71%，孕育了地球上的原始生命，为人们提供了丰富的生产、生活资源和空间资源，是全球生命支持系统的重要组成部分。在全球经济迅速发展和人口激增的情况下，海洋对人类实现可持续发展起到了重要的作用。但随着海洋资源的开发和利用，海洋也受到了严重的污染，其中石油污染表现得尤为突出（图 4-3-11）。

The ocean, accounting for 71% of the Earth's surface area, has gave birth to primitive life on Earth and provided abundant production and living resources and space for people, which makes it an important part of the global life support system. With the rapid development of the global economy and surging population, the ocean has played an important role in the sustainable development of mankind. But with the exploitation and utilization of marine resources, the ocean has also been seriously polluted, especially the oil pollution (Fig. 4-3-11).

海洋石油污染的来源：自然来源约占 92%，人类活动来源约占 8%。造成污染的原因主要包括河流携带来自炼油厂、石油化工厂、油田等工矿企业废水量大，含油浓度高；沿海工业排放；大气输送；船舶污染；海底石油开采。海洋石油污染会对人类健康、环境、生物、水产业、海滨环境等造成破坏。

Among the sources of oil pollution in the ocean, natural sources account for about 92%, and human activities sources account for about 8%. The main causes of oil pollution include the amount of waste water carried by rivers from refineries, petrochemical plants, oil fields and other industrial and mining enterprises is large, the concentration of oil is high; coastal industrial emissions; atmospheric transport; pollution from ships; subsea oil extraction. Marine oil pollution can cause damage to human health, environment, biology, aquaculture, coastal environment and so on.

图 4-3-11　石油污染海洋
Fig. 4-3-11　Oil pollution of the ocean

陆上或浅水井的井喷通常会得到迅速处理，不会释放多少石油，但深水井喷的问题更多。石油的释放量是很多的。与油轮或管道漏油不同，任何一个国家的严重井喷都有很长一段时间。这些灾害的局部性质使人们难以分享自己所学到的知识。科学专家组估计，20世纪90年代，勘探和生产平台的溢漏平均每年向海洋排放约 20 000 t 石油。联合国环境组织派出的一个专家组和联合国粮食及农业组织应墨西哥政府的请求，调查漏油及其影响。图 4-3-12 所示为溢油趋势。

Blowouts from onshore or shallow wells are usually dealt quickly and release little oil, but deepwater blowouts are more problematic because they release large amounts of oil. Unlike oil spills from oil tankers or pipelines, severe blowouts in any country can last for a long time. The localized nature of these disasters makes it difficult for people to share what they have learned. The panel estimated that during the 1990s, oil spills from exploration and production platforms released an average of about 20,000 tonnes of oil into the ocean each year. At the request of the Mexican government, a team of experts from the United Nations Environment and the United Nations Food and Agriculture Organization had been sent to investigate the spill and its effects. Fig. 4-3-12 shows oil spill trend.

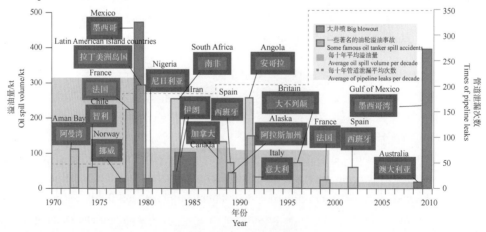

图 4-3-12　溢油趋势
Fig. 4-3-12　Oil spill trend

3.2.2 海洋环境中的石油烃降解菌
3.2.2 Petroleum Hydrocarbon Degrading Bacteria in the Marine Environment

埃克森·瓦尔迪兹灾难和深水地平线钻井平台井喷的石油泄漏凸显了在海上处理石油泄漏的有效生物修复方法的必要性。尽管加强土壤和反应物中石油污染治理的生物修复技术的应用已经有几年的历史，目前还没有实验可以很好地评定碳氢化合物在海洋中的降解程度。

The Exxon Valdez disaster and the deep water horizon blowout have highlighted the need for effective bio-remediation methods to deal with oil spills at sea. Although using bioremediation techniques to enhance controlling oil contamination in soils and reactants have been around for several years, no experiments have been done to assess the extent of hydrocarbon degradation in the ocean.

生物修复是治理海洋石油污染的重要手段，生物修复的大规模应用也是以海洋溢油的治理为开端的。石油污染海洋的生物修复强调自然过程的人工强化，而微生物是其中的工作主体。烃类是天然产物，所以海洋细菌一般都有降解石油的能力，最常见的降解菌有无色杆菌属、黄杆菌属、不动杆菌属、弧菌属、芽孢杆菌属、节杆菌属、诺卡式菌属、棒杆菌属和微球菌属。许多海洋酵母菌和霉菌可以依赖石油和烃类生长，最常见的酵母是假丝酵母属、红酵母属和短梗霉属，霉菌有青霉菌和曲霉菌。另外，藻类和原生动物对修复石油污染也有重要作用。微生物修复石油污染主要有两种：加入具有高效降解能力的菌株和改变环境，促进微生物的代谢能力。在许多情况下，石油生物修复可在现场进行，而对于污染的沉积物，一般使用生物反应器进行治理。生物修复的方法有①添加养分以促进石油降解菌的生长繁殖；②使用分散剂以促进微生物对石油的利用；③接种石油降解菌提高降解效率；④提供电子受体。

Bioremediation is an important means to control marine oil pollution, and the large-scale application of bioremediation also begins with the treatment of marine oil spill. Bioremediation of marine oil pollution emphasizes the artificial enhancement of natural processes, in which microorganisms are the main body of work. Hydrocarbons are natural products, so marine bacteria generally have the ability to degrade petroleum. The most common degrading bacteria are achromobacter, flavobacterium, acinetobacter, vibrio, bacillus, arthrobacter, nocardia, corneobacter, and micrococcus. Many marine yeasts and molds can feed on oil and hydrocarbons. The most common marine yeasts are candida, rhodochroma, and brachyderma. The molds are penicillium and aspergillus. In addition, algae and protozoa play an important role in repairing oil pollution. The main ways of microbial remediation of oil pollution are adding strains with high degradation ability; and changing the environment and promoting the metabolic capacity of microorganisms. In many cases, bioremediation of oil contamination can be performed on-site, and the contamination resulting sediment can generally be treated with bioreactors. Bioremediation methods include ①adding nutrients to promote the growth and reproduction of oil-degrading bacteria; ②using dispersants to promote microbial utilization of oil; ③ inoculating oil-degrading bacteria to improve the degradation efficiency; ④providing electron acceptors.

生物降解是大多数进入海洋环境的石油的最终归宿。在海洋环境中，石油烃降解菌无处不在。实验数据显示微生物可以降解天然油污。然而，生物降解速率较慢，而且对水中和海岸线上的海洋生物有严重的毒性影响。越来越多的数据表明，在石油泄漏的现场条件下参与的特定微生物，一般来说是微生物群。研究的方向应该是为石油烃降解菌的快速生长建立条件。由于这些细菌能降解有毒的多环芳烃，所以适宜的菌株是环菌。微生物降解对油的去除起着重要作用。

Biodegradation is the final destination of most petroleum entering the marine environment. In the marine environment, petroleum hydrocarbon degrading bacteria are ubiquitous. Experiment data showed that microorganisms could degrade natural petroleum pollution. However, biodegradation is slow and has serious toxic effects on marine life in the water and along coastlines. There is a growing number of data indicate that the research direction of specific microorganisms (microbiota) involved in the oil spill site should be the establishment of rapid growth conditions for petroleum hydrocarbon degrading bacteria. Because these bacteria degrade toxic PAHs, the appropriate strains are cyclobacteria. Microbial degradation plays an important role in petroleum removal.

对收集的细胞进行显微镜观察还发现，显性细胞类型呈现出典型的海洋螺旋体的独特形态。图4-3-13所示为1 099~1 219 m的显性细菌（扫描电子显微镜照片）和离源头很远的橘红色斑。

Microscopic examination of the collected cells also revealed that the dominant cell types showed a distinctive pattern of typical marine spirochaetes. Fig. 4-3-13 shows dominant bacteria at 1,099 to 1,219 m and acridine orange stain (inset) with distance from source.

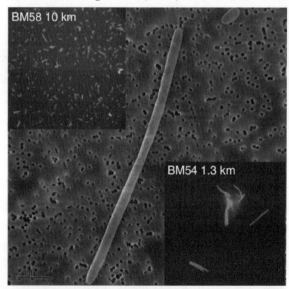

图4-3-13　1 099~1 219 m的显性细菌和离源头很远的橘红色斑（见彩插）

Fig. 4-3-13　Dominant bacteria at 1,099 to 1,219 m and acridine orange stain (inset) with distance from source (see the color figure)

3.2.3 石油烃降解菌的降解机理
3.2.3 Degradation Mechanism of Petroleum Hydrocarbon Degrading Bacteria

图 4-3-14 所示为 LC-烷烃代谢调控机理。该图的顶部为预测的感知蛋白质和烷烃的趋化途径。该图的中部为预测的 LC-烷烃降解路径。该图的底部为 LC-烷烃利用途径。预测的正烷烃转运体在图的左下角和右上角都有标记。实线表示已知的功能途径，未知或不确定的功能途径用虚线表示。被圆圈围绕的加号表示正调节，被圆圈包围的负号表示负调节。OM 为外膜，CM 为细胞质膜（内膜）。

Fig. 4-3-14 shows proposed regulation of LC-alkane metabolism. The top, middle and bottom of the mechanism diagram respectively illustrate the predicted chemotaxis pathways for sensing proteins and alkanes, the predicted regulatory pathways for LC-alkane degradation and the pathways for LC-alkane utilization. Predicted n-alkane transporters are labeled in the bottom left and top right corners of the diagram, with solid lines indicating known functional pathways and dashed lines indicating unknown or uncertain functional pathways. A plus sign surrounded by a circle indicates positive regulation, and a minus sign surrounded by a circle indicates negative regulation. OM is the outer membrane, and CM is the cytoplasmic membrane (inner membrane).

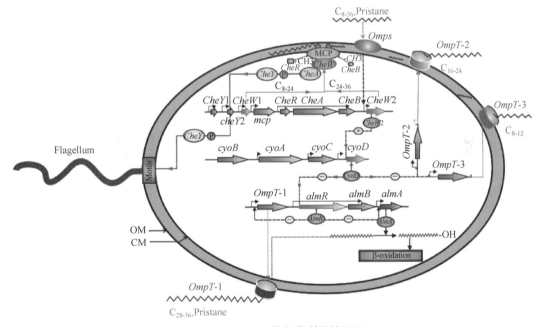

图 4-3-14 LC-烷烃代谢调控机理

Fig. 4-3-14 Proposed regulation of LC-alkane metabolism

机理图中的第一步是通过细胞外烷烃的 *Omps* 的传感，它引发了对烷烃具有特异性的趋化复合蛋白基因的表达。然后，*CheW* 将烷烃刺激传递到调节系统中，并通过一种有待解开的机制诱导 *cyo* 表达。在 LC-烷烃中，*cyo* 的高效表达抑制了 *almR* 的表达，从而抵消了 *almR* 的负面效应。相反，*OmpT*-1 的表达被诱导，启动了 LC-烷烃在外膜上的转运。血浆周围积存的烷烃与 MCP 接触。同时，*almA* 的表达被启动，导致 LC-烷烃的氧化。短链和中链烷烃也有相应的途径，包括 *CheW*1、*OmpT*-2/3 和特定的调节剂。

The first step in the mechanistic diagram is the induction of the expression of chemotactic complex protein genes specific for alkanes by the sensing of Omps for extracellular alkanes. *CheW* then delivers the alkane stimulus to the regulatory system and induces *cyo* expression through a mechanism that remains to be unraveled. In LC-alkanes, the high expression of cyo inhibited *almR* expression, thus counteracting the negative effect of *almR*. In contrast, induction of *OmpT*-1 expression initiates the translocation of LC-alkanes across the outer membrane, bringing the accumulated alkanes around the plasma into contact with the MCP. At the same time, expression of *almA* results in oxidation of LC-alkanes. Corresponding pathways also exist for short-chain and medium-chain alkanes, including *CheW*1, *OmpT*-2/3, and specific modulators.

3.2.4 海洋环境中的石油泄漏污染的修复工程实例
3.2.4 Examples of Restoration Projects for Oil Spill Pollution in the Marine Environment

海洋环境中的石油泄漏污染的修复工程是通过飞机在海面喷洒分散剂展开的。分散剂进入海水中后初始分散，细菌繁殖后会进入石油液滴中。石油和分散剂均会被细菌降解，最后细菌聚集体将会被原生动物和线虫类降解。如图 4-3-15 所示，整个修复过程需要 4 周左右。

Remediation project for oil spill pollution in the marine environment are carried out by spraying dispersants on the surface of the sea from aircraft. When the dispersant enters the sea water, it begins to disperse, and the bacteria multiply into the oil droplets, where both the oil and dispersant are degraded by the bacteria. Eventually, the bacterial aggregates will be degraded by protozoa and nematodes. As shown in Fig. 4-3-15, the entire remediation process will take about four weeks.

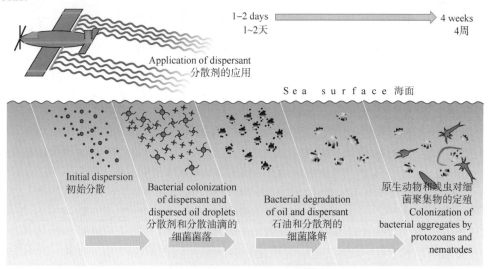

图 4-3-15 石油污染修复过程
Fig. 4-3-15 Oil pollution remediation process

3.3 可降解塑料的生物生产及合成塑料的生物降解
3.3 Bioproduction of Biodegradable Plastics and Biodegradation of Synthetic Plastics

塑料产品的广泛使用给人们的生产、生活带来了革命性的变化，并产生了巨大的经济效益和社会效益。传统的塑料制品最主要的缺陷就是不具有降解性，因此塑料制品的不断老化、废弃、丢弃，对环境的污染也日益加重。人们已经研究开发或改进了多种处理方法，但每种方法都有其缺点。因此，环保型材料的使用仍是一个亟待解决的问题。

The widespread use of plastic products has brought revolutionary changes to production and human life, and has produced huge economic and social benefits. The main defect of traditional plastic products is that they are not degradable. Therefore, the continuous aging, abandonment and discarding of plastic products lead to increasingly serious environment pollution. Many treatments have been developed or improved, but each has its drawbacks. Therefore, the use of environment-friendly materials is still an urgent problem to be solved.

3.3.1 可降解塑料的种类
3.3.1 Types of Biodegradable Plastics

可降解塑料 Degradable plastics
- 生物降解塑料（Biodegradable plastics）
- 光降解塑料（Photodegradable plastics）
- 化学降解塑料（Chemically degradable plastics）
- 光/氧生物复合降解塑料（Photo/Oxygen biodegradable plastics）

3.3.2 可降解塑料的生物生产
3.3.2 Bioproduction of Biodegradable Plastics

微生物通过生命活动可合成高分子，主要包括微生物聚酯和微生物多糖，其中微生物聚酯方面的研究较多。微生物聚酯是一种耐紫外辐射、具有生物降解性和热塑性、易加工的生物高分子，其具有良好的生物降解性和生物组织相容性。

Microorganisms can synthesize polymers through life activities, mainly microbial polyester and microbial polysaccharide, among which the former has been studied more. Microbial polyester is a kind of ultraviolet radiation resistant, bio-degradable, thermoplastic and easy-to-process bio-polymer with good biodegradability and biohistocompatibility.

A. PHA 的生物合成

A. Biosynthesis of PHA

聚羟基脂肪酸酯（PHA）是由很多细菌合成的一种胞内聚酯，在生物体内主要是作为碳源和能源的贮藏性物质而存在的。与很多生物基塑料相比（如 PLA、PBS）PHA 完全是由生物合成和生物聚合的。目前蓝晶生物科技有限责任公司（简称蓝晶公司）PHA 生物合成技术比较成熟，如图 4-3-16 所示。

Polyhydroxyfatty acid ester (PHA) is an intracellular polyester synthesized by many bacteria, which is mainly used as carbon source and energy storage substance in organism. Compared with many bio-based plastics (such as PLA and PBS), PHA is completely biosynthesized and biopolymerized. At present, the PHA biosynthesis (Fig. 4-3-16) technology of Bluepha Biotechnology Co. Ltd. referred to as Bluepha Company is relatively mature.

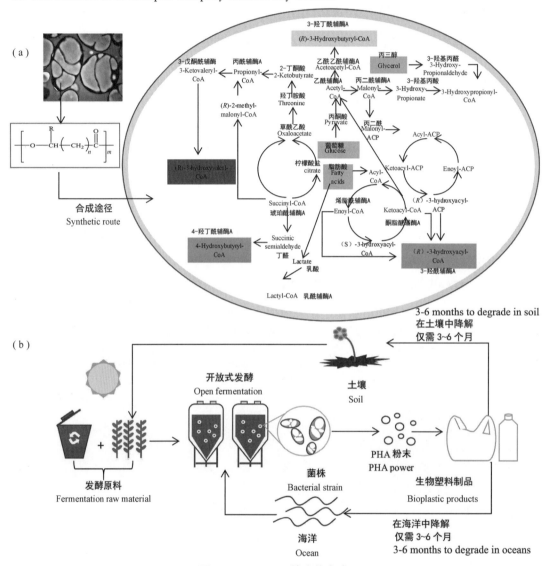

图 4-3-16 PHA 的生物合成

Fig. 4-3-16 Biosynthesis of PHA

(a) PHA 细胞内合成途径；(b) 蓝晶公司 PHA 生产流程

(a) PHA intracellular synthesis pathway; (b) PHA production process of Bluepha company

B. PHB 的生物合成

B. Biosynthesis of PHB

聚 3-羟基丁酸酯（PHB）是 PHA 中存在最广、发现最早且研究也最为透彻的一种。普通的 PHB 均聚物非常硬且脆，工业上使用加工存在较大问题。因此，通过调控微生物碳源，成功开展了具有各种特性的共聚体聚酯研究。在 3HB 的各种共聚物中，3-羟基丁酸酯与 3-羟基戊酸酯的共聚物 PHBV 的研究最多。日本的土肥小组，在 A. eutrophus 中加入 ^{13}C 标识的丙酸，并用 NMR 方法判断了 ^{13}C 所引入共聚体聚酯的位置，提出了 P（3HB-co-3HV）生物合成途径，如图 4-3-17 所示。

Poly-3-hydroxybutyrate is one of the most widely existed, the earliest discovered, and most thoroughly studied PHA. Ordinary PHB homopolymers are very hard and brittle, which has great problems in industrial application and processing. Therefore, by regulating microbial carbon sources, various characteristics of copolymerization polyester were successfully studied. Among the various copolymers of 3HB, the copolymer PHBV of 3-hydroxybutyrate and 3-hydroxyvalerate has been studied most. The Japanese soil fertilizer group added ^{13}C-labeled propionic acid to A. eutrophus, and determined the position of the copolymer polyester introduced by ^{13}C by NMR method, and proposed the P（3HB-co-3HV）biosynthesis pathway, as shown in Fig. 4-3-17.

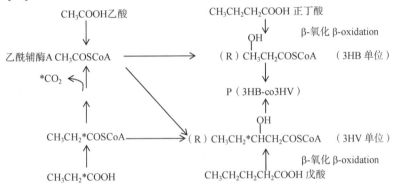

图 4-3-17　PHBV 的生物合成途径

Fig. 4-3-17　Biosynthetic pathway of PHBV

3.3.3　合成塑料的生物降解
3.3.3　Biodegradation of Synthetic Plastics

生物降解聚合物通常经历两步：初步降解和最终的生物降解。首先聚合物主链断裂形成低相对分子质量的可以被微生物同化的碎片，然后这些碎片进入微生物细胞中，进一步发生同化作用转化为二氧化碳、水等产物。图 4-3-18 所示为生物降解聚合物的生物降解过程。

Biodegradable polymers usually undergo two steps, initial degradation and final biodegradation. Firstly, polymerize backbone breaks to form relatively low molecular mass fragments that can be assimilated by microorganisms. These fragments then enter microbial cells, where further assimilation occurs and they are converted into products such as carbon dioxide and water. Fig. 4-3-18 shows biodegradation process of biodegradable polymers.

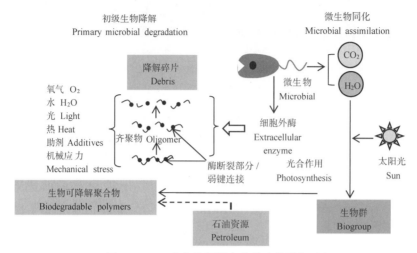

图 4-3-18 生物降解聚合物的生物降解过程

Fig. 4-3-18 Biodegradation process of biodegradable polymers

A. 聚苯乙烯的生物降解

A. Biodegradation of polystyrene

塑料在环境中难以自然降解，而聚苯乙烯（PS）又是其中之最。由于高分子量和高稳定性，普遍认为微生物无法降解聚苯乙烯类塑料。2015年北京航空航天大学杨军教授研究组证明了黄粉虫（面包虫）的幼虫可降解聚苯乙烯这类最难降解的塑料（图4-3-19）。该突破性研究显示，以聚苯乙烯泡沫塑料作为唯一食源，黄粉虫幼虫可存活1个月以上，最后发育成成虫，其所啃食的聚苯乙烯被完全降解矿化为 CO_2 或同化为虫体脂肪。

Plastics are difficult to degrade naturally in the environment, and polystyrene (PS) is the most difficult among them. Due to its high molecular mass and high stability, it is generally believed that microorganisms cannot degrade polystyrene plastics. In 2015, Professor Yang Jun and his team at Beijing University of Aeronautics and Astronautics proved that the larvae of the yellow mealworm (bread worm) can degrade polystyrene, the most difficult plastic to degrade (Fig. 4-3-19). This breakthrough study showed that mealworm larvae could survive for more than one month with polystyrene foam as the only food source and eventually develop into adult worms. The polystyrene they eat is completely degraded and mineralized into CO_2 or assimilated into insect fat.

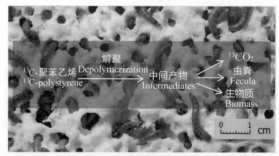

图 4-3-19 黄粉虫啃食聚苯乙烯（见彩插）

Fig. 4-3-19 Yellow mealworms are eating polystyrene (see the color figure)

B. PET 的生物降解

B. Biodegradation of PET

对苯二甲酸乙二酯（PET）是一种常见的芳香聚酯，长久以来 PET 被认为基本不能被微生物降解（发现的 PET 降解微生物局限于几种真菌如 *Fusarium oxysporum* 和 *F. solani*）。可喜的是，2016 年 Shosuke Yoshida 等人分离到了一株 PET 降解菌 *Ideonella sakaiensis* 201-F6，它能够利用 PET 作为唯一的碳源和能源，并对其的降解机制进行了预测（图 4-3-20）。

Ethylene terephthalate (PET) is a common aromatic polyester, and for a long time PET was considered to be basically unable to be degraded by microorganisms (the discovered PET degradation microorganisms found were limited to several fungi such as *Fusarium oxysporum and Fusarium solani*). Fortunately, a PET degrading bacterium *Ideonella sakaiensis* 201-F6 isolated by Shosuke Yoshida et al. (Fig. 4-3-20). In 2016 was able to use PET as the only carbon source and energy, and its degradation mechanism was predicted.

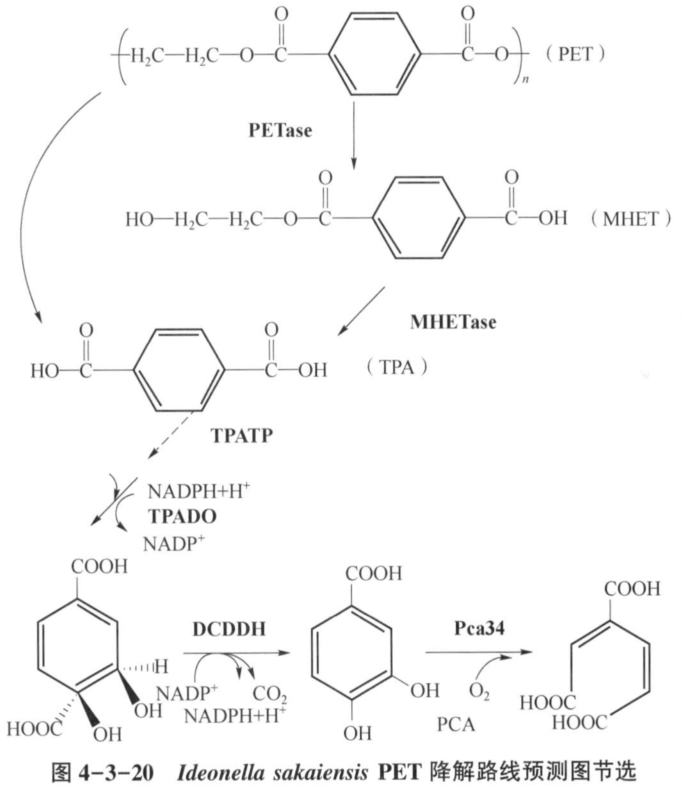

图 4-3-20　*Ideonella sakaiensis* PET 降解路线预测图节选

Fig. 4-3-20　Excerpt from the prediction map of PET degradation pathway of sakaiensis

注释：

胞外 PET 解聚酶水解 PET 产生 MHET（主要产物）和 TPA。MHET 酶（一种预测的脂蛋白）水解 MHET 产生 TPA 和 EG，TPA 结合 TPA 运载体（TPATP）后被 TPA 1,2-双加氧酶（TPADO）和 DCDDH（一种脱氢酶）作用产生 PCA，接着 PCA 被 PCA 3,4-双加氧酶继续降解。

Note：

Extracellular PET depolymerase hydrolyzes PET to produce MHET and TPA. MHET enzyme (a predicted lipoprotein) hydrolyzes MHET to produce TPA and EG, and TPA binds TPA carrier (TPATP) to produce PCA by TPA 1, 2-dioxygenase (TPADO) and DCDDH (a dehydrogenase), which is then further degraded by PCA 3, 4-dioxygenase.

C. 聚己内酯的生物降解

C. Biodegradation of PCL

聚己内酯（PCL）由己内酯开环聚合反应制备得到，具有生物降解性能。如图 4-3-21 所示，降解 PCL 的酶主要是脂肪酶、酯酶和角质酶。PCL 降解时，聚合物链首先发生水解生成 6-羟基正己酸，然后生成 ω-氧化中间产物、β-氧化中间产物和乙酰-CoA，后者在 TCA 循环中可以进一步发生降解。PCL 的降解首先发生在无定形区域，随着降解的发生，PCL 的结晶度提高。有趣的是，PCL 与植物表皮的角质聚酯结构是类似物，这与 PCL 能被植物致病真菌中的角质酶促降解相关联。

Polycaprolactone (PCL) is prepared from cyclo-opening polymerization of caprolactone and has biodegradability. As shown in Fig. 4-3-21, the main enzymes that degrade PCL are lipase, esterase, and cutinase. When PCL is degraded, the polymer chain is firstly hydrolyzed to 6-hydroxy-n-caproic acid, then to ω-oxidation intermediate, β-oxidation intermediate and acetyl-CoA, which can be further degraded in the TCA cycle. The degradation of PCL first occurred in the amorphous region, and with the occurrence of degradation, the crystallinity of PCL increased. Interestingly, PCL is an analogue of the keratin polyester structure of the plant epidermis, which correlates to the fact that PCL can be degraded by keratin enzymes in plant-pathogenic fungi.

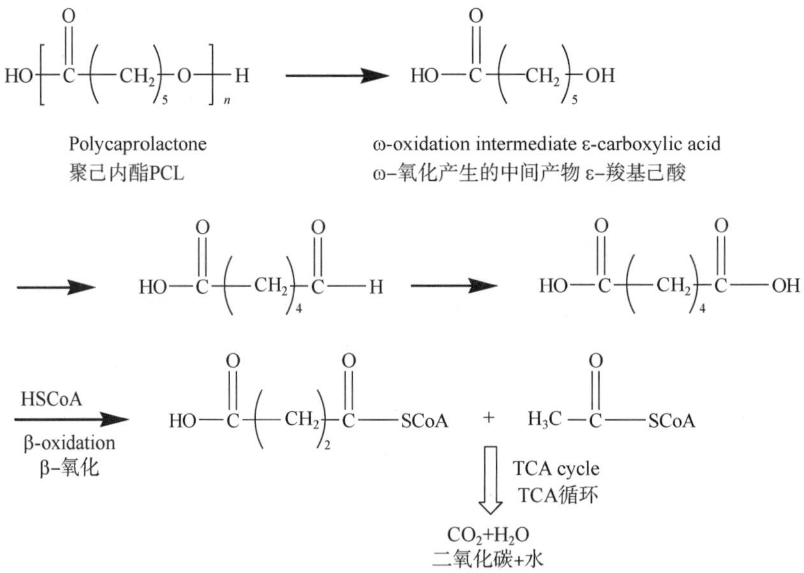

图 4-3-21 PCL 的降解机理

Fig. 4-3-21 Degradation mechanism of PCL

D. PLA 的生物降解

D. Biodegradation of PLA

聚乳酸（PLA）是由可再生资源——乳酸聚合而成的高分子聚酯，是一种具有潜力的绿色生物塑料。然而，自然界中已知的降解 PLA 的微生物并不多，已发现的大部分降解菌属于放线菌。微生物分泌胞外解聚酶，解聚酶不能穿透 PLA 基质，酶促降解仅发生在 PLA 表面，随着 PLA 的降解，酶扩散进入 PLA 的无定形区，随后 PLA 的结晶区域被降解。解聚酶的降解使 PLA 分子的酯键断裂，产生寡聚体、二聚体和单体，最终分解成为 CO_2 和 H_2O。

Polylactic acid (PLA) is a high-molecular polyester polymerized by lactic acid, a renewable resource, which makes it to be a potential green bioplastic. However, few microorganisms are known to degrade PLA in nature, and most of them are actinomycetes. Microorganisms secrete extracellular depolymerase that cannot penetrate the PLA matrix, so the enzymatic degradation reaction only occurs on the PLA surface. With the degradation of PLA, the enzyme diffused into the amorphous region of PLA, and subsequently the PLA crystalline region is degraded. Degradation of depolymerase breaks the ester bond of PLA molecule, producing oligomers, dimers and monomers, which eventually decompose into CO_2 and H_2O.

3.4 厌氧氨氧化及废水处理的应用
3.4 Application of Anaerobic Ammonium Oxidation and Waste Water Treatment

氮含量的控制是水质检测工作的重要组成部分。随着工业社会的不断发展，水体的富营养化已成为常态。当前，对于氮污染处理，人们在传统处理程序中选择的是硝化-反硝化来完成脱氮，在这个过程中，需要使用碱和碳源，这会造成二次污染，也会形成更为高昂的成本。因此，运用现代科技来对现有的脱氮技术进行优化，已成为人们关注的重点。厌氧氨氧化技术能够在保持低成本的基础上来提升能效，有着极高的研究价值。

Nitrogen content control is an important part of water quality testing. With the development of industrial society, eutrophication of water has become normal. At present, for nitrogen pollution, the traditional treatment, nitrification-denitrification, is often used to remove nitrogen. In the process, alkali and carbon sources are used, which causes secondary pollution and is also more expensive. Therefore, the application of modern technology to optimize the existing nitrogen removal technology has become the focus of attention. In particular, anammox technology can improve energy efficiency with low cost, which has excellent research value.

3.4.1 厌氧氨氧化反应原理
3.4.1 Mechanism of Anammox

在厌氧环境下，Anammox 菌能够以 NO_2^-—N 为电子受体，以 NH_4^+—N 为电子供体，使二者反应生成 N_2。Anammox 菌的主要代谢途径：首先，Cyt cd1 型亚硝酸还原酶（NIR）将 NO_2^- 还原成 NO；随后，在联氨水解酶（HH）的作用下，NO 和 NH_4^+ 缩合成为 N_2H_4；联氨氧还酶（HAO）将生成的 N_2H_4 氧化为 N_2；与此同时，亚硝酸氧化酶（Nar）将 NO_2^- 氧化成 NO_3^-。以上代谢过程在细菌内部完成，由 N_2H_4 作为供体所释放的 4 个电子，通过细胞色素 C、辅酶 Q 等进行传递，最终抵达受体 NIR 和 HH（4 个电子中 1 个电子交给 NIR，3 个电子交给 HH）。随着电子的传递，质子被排至厌氧氨氧化体膜的外侧，并在膜两侧形成质子梯度,从而驱动 ATP 和 NADPH 的合成。图 4-3-22 所示为厌氧氨氧化原理。

Under anaerobic conditions, Anammox can use NO_2^-—N as an electron acceptor and NH_4^+—N as an electron donor to react with each other to form N_2. The main metabolic pathways of anammox: First, Cyt cd1 nitrite reductase (NIR) reduces NO_2^- to NO; then, under the action of hydrazine hydrolase (HH), NO and NH_4^+ are condensed into N_2H_4; hydrazine oxygen-reducing enzyme (HAO) oxidizes the produced N_2H_4 to N_2; at the same time, nitrite oxidase (Nar) oxidizes NO_2^- to NO_3^-. The above metabolic process is completed inside the bacteria, and the four electrons released by N_2H_4 as donors are transmitted by cytochrome C, coenzyme Q, etc., and finally reach the receptors NIR and HH (1 electron of 4 electrons is given to NIR, 3 electrons are given to HH). As electrons are transferred, protons are discharged to the outside of the anammox membrane, and proton gradients are formed on both sides of the membrane, thereby driving the synthesis of ATP and NADPH. Fig. 4-3-22 shows mechanism of anammox.

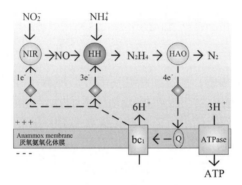

图 4-3-22 厌氧氨氧化原理

Fig. 4-3-22 Mechanism of anammox

3.4.2 厌氧氨氧化菌
3.4.2 Anammox Bacteria

厌氧氨氧化菌(图 4-3-23)属于分支很深的浮霉菌纲,已发现并鉴定的厌氧氨氧化菌共有 6 属 18 种,构成了独立的厌氧氨氧化菌科(*Anammoxaceae*)。其中,大多数细菌从污水处理厂或实验室反应器内得到,也有少数来自海水样品。厌氧氨氧化菌倍增时间很长(11 天),并且迄今为止仍未获得其纯培养物。正因如此,基于纯培养物分离所开展的形态、代谢、生化和遗传等方面的传统分析方法,在应用于厌氧氨氧化菌反应器中微生物的种类、丰度及相互之间的关系的研究时受到了很大的限制。

The anammox bacteria (Fig. 4-3-23) belongs to the deeply branched planctomycetia. At present, the independent anammoxaceae family consists of 18 species in 6 genera that have been discovered and identified. Most of the bacteria came from sewage plants or laboratory reactors, but a few came from seawater samples. Anammox bacteria has a long doubling time (11 days) and no pure cultures have been obtained yet. Because of this, the traditional analytical method of morphological, metabolic, biochemical and genetic based on pure culture isolation are very limited when applied to the study of microbial species, abundance and relationships in anammox bacteria reactors.

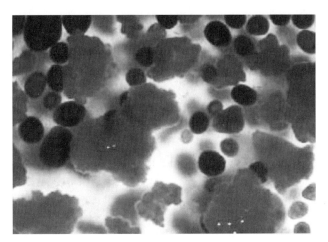

图 4-3-23 厌氧氨氧化菌
Fig. 4-3-23 Anammox bacteria

3.4.3 厌氧氨氧化污水处理工艺
3.4.3 Ananmmox Waste Water Treatment Process

A. 全自氧脱氨工艺

A. The Whole Auto-oxygen Deamination Process

利用全自氧脱氨工艺进行污水处理时,主要通过控制溶解氧实现亚硝化和厌氧氨氧化,并且在这个污水处理过程中,自养菌将水体中的氮元素以及氨元素转化为氮气。在进行污水处理时,整个处理过程都是在微好氧的环境下进行的,通过亚硝化菌化学反应生成亚硝氮,亚硝氮在与剩下的氮氨厌氧氨氧化反应生成氮气。由于亚硝氮菌和厌氧氨氧化菌都属于自养型细菌的范畴,因此,在进行全自氧脱氨工艺污水处理时,不需要添加外源有机物,只需在无机自养环境下进行即可。

In the full auto-oxygen deamination process of waste water treatment, nitrosation and anammox oxidation are achieved mainly by controlling dissolved oxygen. In the whole process, nitrogen and ammonia in the water are converted into nitrogen by autotrophic bacteria in the microaerobic environment. Nitrite is generated by chemical reaction of nitrite bacteria, and nitrite reacts with the remaining ammonia and anammox to produce nitrogen. Since nitrite bacteria and anammox bacteria belong to the category of autotrophic bacteria, it is not necessary to add exogenous organic matter in the waste water treatment of the total autoaerobic deamination process, but only in the inorganic autotrophic environment.

B. 亚硝化厌氧氨氧化工艺

B. The Nitrosation Anaerobic Oxidation Process

亚硝化厌氧氨氧化工艺是如今污水处理中最常用的一种工艺,在进行污水处理时,它主要分为两个阶段，这两个阶段在不同的容器中进行反应。第一阶段是亚硝化阶段，该阶段可以将污水中 50％左右的氨元素与氮元素转化为亚硝态氨；第二阶段是厌氧氨氧化阶段，该阶段将污水中剩余的氨元素与氮元素以及转化生产的亚硝态氨厌氧氨氧化反应，转变为氨气，从而实现脱氮的目的。亚硝化厌氧氨氧化工艺具有三大优点：首先是通过亚硝化厌氧氨氧化工艺生成亚硝态氨，这种物质属于碱性物质，而厌氧水中已产生一些重碳酸盐，这样就实现了酸碱中合，有助于实现水体平衡；其次在进行亚硝化厌氧氨氧化污水处理时，在不同的容器中进行反应处理，反应容器环境的不同，为功能菌提供了更为适合自身的生长环境，这样可以减少进水物质对厌氧氨氧化菌的抑制作用；最后，利用亚硝化厌氧氨氧化工艺进行污水处理。

Nitrosation anammox is one of the most commonly used ammox processes in waste water treatment. When sewage treatment is carried out, it is mainly divided into two stages of reaction in different containers. The first stage is nitrification, which can convert about 50% of ammonia and nitrogen elements in sewage into nitrite ammonia. The second is the anaerobic ammox stage, in which the remaining ammonia element in the sewage oxidizes with nitrogen element, nitrite ammonia and anaerobic ammonia, and converts into ammonia gas, so as to achieve the purpose of nitrogen removal. Nitrification and anammox process has three advantages: first, because nitrification and anammox process can generate basic material nitrite ammonia, and anaerobic water has produced some bicarbonate, water can achieve balance by acid-base synthesis; secondly, nitrification and anammox waste water treatment is carried out in different containers which provide functional bacteria with more suitable growth environment choices, so as to reduce the inhibitory effect of influent substances on anammox bacteria. Finally, the nitrosation anammoxidation process was used for waste water treatment.

3.4.4 实际工程应用
3.4.4 Practical Engineering Applications

随着厌氧氨氧化水处理工程的应用，研究不断取得突破，实际工程的建设在全世界范围内兴起。据报道，截至 2017 年，全世界厌氧氨氧化工程已超过 110 座。上述已建或在建厌氧氨氧化工程应用于污水处理，因消化液具有温度高、水质波动小、低碳氮比的特点，世界上第一座厌氧氨氧化工程就是用于处理污泥消化液，世界上半数以上的工程也是用于处理污泥消化液。

With the application of anammox water treatment engineering, the research on this technology has also made continuous breakthroughs, and the construction of practical projects is rising in the world. It is reported that up to 2017, there have been more than 110 anammox projects in the world. The above-mentioned anaerobic ammonia oxidation project has been built or under construction for sewage treatment. Because the digestive liquid has the characteristics of high temperature, small water quality and low carbon-nitrogen ratio, the world's first anaerobic ammonia oxidation project is used to treat sludge. Digestive fluid, more than half of the world's projects are also used to treat sludge digestive fluid.

由曹业始博士主导的樟宜水处理系列项目，实现了主流部分硝化和厌氧氨氧化，同时强化了生物除磷,如图 4-3-24 所示。樟宜污水厂采用分段式活性污泥工艺,设计处理量为 80×10^4 m³/d,共 4 个相同的部分,每部分设计处理量为 20×10^4 m³/d,出水水质良好。优势菌种为 AOB 以及 *candidatus brocadia sp.* 40 为主的悬浮的游离 AnAOB,其自养脱氮过程的贡献率为 37.5%。此外,樟宜污水处理厂能耗降低达 60%,产泥率降低达 80%。樟宜污水处理厂主流厌氧氨氧化工程的成功运行与新加坡得天独厚的水温(28~32 ℃)是分不开的,未来主流厌氧氨氧化在低温地区的应用还有很大的研究空间。

The Zhangyi Water Treatment project led by Dr. Cao Yeshi achieved mainstream nitrification and anammox, while enhancing biological phosphorus removal, as shown in Fig. 4-3-24. The Zhangyi Waste Water Treament Plant adopts the segmental activated sludge process (SFAS), with a designed treatment capacity of 80×10^4 m³/d, and a total of four identical parts, each of which has a designed treatment capacity of 20×10^4 m³/d, and the effluent quality is good. The dominant strains are AOB and the suspended free AnAOB mainly composed of *Candida brocadia sp.* 40, whose contribution rate of autotrophic nitrogen removal process is 37.5%. In addition, the Zhangyi Waste Water Treatment Plant reduced energy consumption by 60% and sludge yield by 80%. The successful operation of the mainstream anammox project in Zhangyi Waste Water Treatment Plant is inseparable from Singapore's unique water temperature (28-32 ℃). There is much room for future research on the application of mainstream anammox in low-temperature regions.

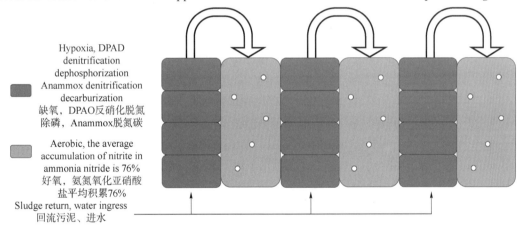

图 4-3-24 樟宜污水处理厂工作流程

Fig. 4-3-24 Workflow of the Zhangyi Waste Water Treament Plant

3.5 合成生物学在环境生物技术中的应用
3.5 Application of Synthetic Biology in Environmental Biotechnology

在美国、印度和中国在内的工业化国家，不断增加的空气、土壤、地表和地表水污染对公共健康和生态环境构成了重大威胁。影响土壤和水的大多数污染物是重金属和有机化合物，如矿物油烃、多环芳烃、苯衍生物和卤代烃。许多用于农业的有机污染物（农药二氯二苯基三氯乙烷、莠去津和五氯苯酚）、工业有机污染物（溶剂如二氯乙烷或介电液体如多氯联苯）或军事用途的有机污染物（爆炸物如2，4，6-三硝基甲苯）是人为起源的异生素。尽管这些污染物具有较强的顽固性，但许多化合物或多或少地可进行生物降解。除了这些环境恶化的传统原因之外，近几十年来人类的二氧化碳和其他温室气体排放量逐渐增加，随之而来的是对气候的影响。

Public health and the ecological environment in the United States, India, and China are seriously threatened by increasing air, soil, surface, and surface water pollution. The major pollutants affecting soil and water are heavy metals and organic compounds such as mineral oil hydrocarbons, polycyclic aromatic hydrocarbons, benzene derivatives, and halogenated hydrocarbons. Many organic pollutants used in agriculture (the pesticides dichlorodiphenyltrichloroethane, atrazine, and PCP), industrial organic pollutants (solvents such as dichloroethane or dielectric liquids such as polychlorinated biphenyls), or militaryorganic pollutants (explosives such as 2 coronates 4 coronates 6-trinitrotoluene) are heterogenins of anthropogenic origin. Despite their strong recalcitrance, many are more or less biodegradable. In addition to these traditional causes of environmental degradation, the amount of carbon dioxide and other greenhouse gases emitted by humans has gradually increased in recent decades, with the attendant effects on the climate.

微生物及其代谢途径是造成生物圈大规模转化的主要原因。微生物在需氧或厌氧条件下通过完全矿化或共代谢降解有毒化学物质。其有利的特性，如较小的基因组大小、相对简单的细胞、短的复制时间、快速进化和适应新的环境条件使微生物特别是细菌成为生物修复技术的有利候选者，即原生或非原生去除来自环境的污染化学品使用生物制剂。消除工业生产活动造成的环境污染是一个引起生物技术专家关注的重要课题。事实上，工业生产中积累的废旧化学品的价值化是循环经济和第四次工业革命的支柱之一。

Microorganisms and their metabolic pathways are the main causes of large-scale transformation of the biosphere. Microorganisms can degrade toxic chemicals by complete mineralization orcometabolism under aerobic or anaerobic conditions. Microorganisms, especially bacteria, are favorable candidates for bioremediation technologies due to their favorable properties, such as small genome size, simple cell, short replication time, and rapid evolution and adaptation to new environmental conditions. The elimination of environmental pollution caused by industrial production activities is an important topic that has attracted the attention of biotechnologists. In fact, the value of waste chemicals accumulated in industrial production is one of the pillars of the circular economy and the fourth industrial revolution.

合成生物学的一个关键目标是通过结合高通量技术和计算方法来表征和预测细胞行为，从而获得对活细胞的全面、定量的理解。代谢工程，最初被 Bailey 定义为新的科学学科，现在被广泛地认为是优化细胞内遗传和调节过程的重要方法，具有提高生物合成的天然产物的产量和生产力，建立新宿主细胞产物的合成，扩大底物的范围或改善底物的摄取，以及提高整体细胞的稳健性等技术优势。这些目标可以通过在宿主细胞中设计天然代谢途径或由源自不同生物的酶组装的合成途径来实现。合成生物学的目标完全符合代谢工程的目标。这两个学科现在是不可分割、相互补充的，甚至已经融入系统代谢工程领域了。

One key objective of synthetic biology is to obtain a comprehensive, quantitative understanding of living cells by combining high-throughput techniques and computational methods to characterize and predict cell behavior. Metabolic engineering, originally defined as a new scientific discipline by Bailey, is now widely recognized as an important approach for optimizing genetic and regulatory processes within cells. It has the technical advantages of increasing the yield and productivity of biosynthesized natural products, establishing the synthesis of new host cell products, expanding the range of substrates or improving substrate uptake, and improving overall cell robustness. These goals can be achieved by designing native metabolic pathways in host cells or synthetic pathways assembled by enzymes derived from different organisms. Synthetic biology is well aligned with the goals of metabolic engineering, and the two disciplines are now inseparable and complementary to each other, and even integrated into the field of systems metabolic engineering.

如图4-3-25所示，应用于环境生物技术的合成生物学实验流程主要包括以下三个步骤。

第一步：了解污染物，查找和识别相关的反应途径。首先要利用数据库或其他预测系统判断污染物的种类和相关代谢途径；随后确定酶或代谢产物等代谢途径的组成部分；最终通过代谢组学、蛋白质组学等分析技术完成对反应途径的检测与定量分析。

第二步：了解和选择宿主。首先要从天然菌株或实验中保存菌株中选定宿主菌；随后通过数据库或测序技术获得其基因组序列和全基因组代谢模型；最终解析宿主菌生长优势的来源。

第三步：建立和优化宿主内的反应途径。首先通过全基因组代谢模型预测宿主细胞的代谢流量并找到限速步骤，通过基因工程手段调控代谢过程；随后在基因表达水平上调节代谢动力和平衡基因表达；最终通过蛋白质工程提升相关酶活性。

As shown in Fig. 4-3-25, the experimental process of synthetic biology applied to environmental biotechnology mainly includes the following three steps:

Step 1: Understand the contaminant, and find and identify the relevant reaction pathways. First, the types of pollutants and related metabolic pathways should be judged using databases or other prediction systems. Then, the components of metabolic pathways such as enzymes or metabolites are determined. Finally, metabolomics, proteomics and other analysis techniques are used to complete the detection and quantitative analysis of the reaction pathways.

Step 2: Learn and select the host. The first is to select the host bacteria from natural or experimentally preserved strains. Then, the genome sequence and whole genome metabolic model are obtained by database or sequencing technology. Finally, the source of the growth advantage of host bacteria is analyzed.

Step 3: Establish and optimize the response pathways within the host. Firstly, a genome-wide metabolic model is used to predict the metabolic flow of host cells and find the rate-limiting steps, and the metabolic process is regulated by genetic engineering. This is followed by regulation of metabolic motility and balanced gene expression at the gene expression level. Finally, the activities of related enzymes are enhanced by protein engineering.

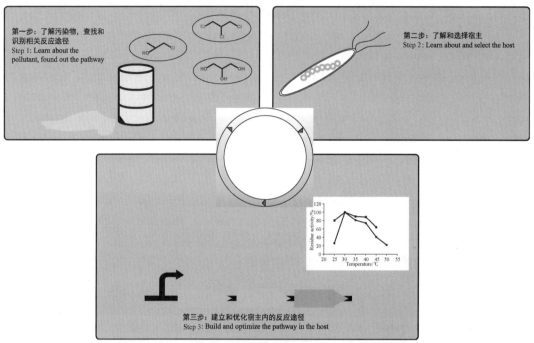

图 4-3-25　应用于环境生物技术的合成生物学实验流程

Fig. 4-3-25　Process of synthetic biology applied for environmental biotechnology

以下是一些合成生物学方法应用于工程生物降解途径和全细胞降解物的实例。

Here are some examples of synthetic biology approaches applied to engineering biodegradation pathways and whole-cell degraders.

Martínez 及其同事将用于从芳香杂环中除硫的 4S 途径分成两个模块（dszC1-D1 和 dszB1-A1-D1），如图 4-3-26 所示，这两个模块在两个恶臭假单胞菌 KT2440 菌株中单独表达以形成合成的联合体。通过将菌株以 1∶4 的比例组合并将它们与 dszB 生产菌株的无细胞提取物混合，使二苯并噻吩在实验条件下几乎完全转化为无硫的 2-羟基联苯。

Martinez and his colleagues split the 4S pathway for sulfur removal from aromatic heterocyclic rings into two modules (dszC1-D1 and dszB1-A1-D1), as shown in Fig. 4-3-26, that were expressed individually in two *P. putida* KT2440 strains to form a synthetic consortium. By combining the strains at a ratio of 1∶4 and mixing them with cell-free extracts of dszB-producing strains, an almost complete conversion of dibenzothiophene to sulfur-free 2-hydroxybiphenyl was achieved under the experimental conditions.

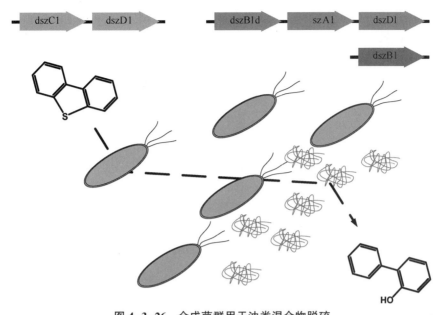

图 4-3-26 合成菌群用于油类混合物脱硫

Fig. 4-3-26 Synthetic bacterial consortium for desulfurization of oil compounds

Wang 等人构建了一个合成细菌聚生体（图 4-3-27），其中两个大肠杆菌菌株中的每个都作为双输入传感器起作用，具有用于检测砷、汞和铜离子以及群体感应分子的合成 AND 门（3OC$_6$HSL）。菌株 2 仅在培养物中存在所有三种金属离子时产生红色荧光。

Wang et al. constructed a synthetic bacterial consortium (Fig. 4-3-27) in which each of the two *E. coli* strains functions as a dual-input sensor. The bacterial consortium has a combined AND gate (3OC$_6$HSL) for the detection of As, Hg, Cu ions and quorum sensing molecules. Strain 2 generated red fluorescence only in the presence of all three metal ions in culture.

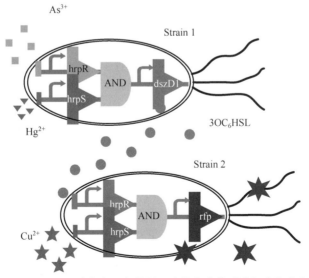

图 4-3-27 通过含有正交基因回路的合成菌群进行生物感应

Fig. 4-3-27 Biosensing through synthetic bacterial consortium with orthogonal genetic circuits

Benedetti 及其同事通过在来自约翰不动杆菌的环己酮响应的 ChnR/P$_{chnB}$ 控制下用来自大肠杆菌的 yedQ 二鸟苷酸环化酶基因组成的正交遗传装置控制环状 di-GMP 水平，实现了恶臭假单胞菌 KT2440 的浮游和生物膜形态的定向转换（图 4-3-28）。设计后的细胞具有用于 1-氯丁烷生物降解的合成操纵子，可形成可诱导的生物膜，具有比游离浮游细菌更高的脱卤素酶活性。

Benedetti and his colleagues used an orthogonal genetic apparatus consisting of the yedQ diguannylate cyclase gene from *E. coli* to control circular di-GMP levels under the control of a cyclohexanone-responsive ChnR/P$_{chnB}$ regulatory node from *Acinetobacter johnsonii* (Fig. 4-3-28). The directional conversion of planktonic and biofilm morphology of *P. putida* KT2440 was achieved. The cells were designed with synthetic operons for 1-chlorobutane biodegradation to form inducible biofilms with higher dehalogenase activity than free planktonic bacteria.

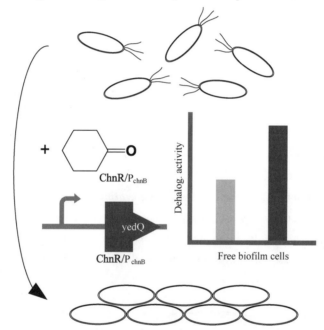

图 4-3-28 通过可调控的 *P. putida* 生物膜提高含氯化合物的生物降解
Fig. 4-3-28 Customizable *P. putida* biofilm enhances biodegradation of chlorinated compounds

Dvorak 及其同事通过结合工程化的卤代烷脱卤素酶 DhaA31，卤代醇脱卤素酶 HheC 和环氧化物水解酶 EchA，固定的交联酶聚集体（CLEAs）和聚乙烯醇颗粒（LentiKats），组装了第一个 1，2，3-三氯丙烷（TCP）的体外合成生物降解途径（图 4-3-29）。该固定化途径在台式填充床反应器中连续工作两个月，可将高浓度并且有毒的 TCP 转化为污染水中的甘油。

Dvorak and his colleagues assembled the 1st *in vitro* synthetic biodegradation pathway for 1, 2, 3-trichloropropane (TCP) by combining engineered haloalkane dehalogenase DhaA31, halohydrin dehalogenase HheC, and epoxide hydrolase EchA, immobilized in cross-linked enzyme aggregates (CLEAs) and polyvinyl alcohol particles (LentiKats) (Fig. 4-3-29). The immobilized pathway converted high concentrations of toxic TCP into glycerol in contaminated water for more than two months of continuous operation in a bench-top packed bed reactor.

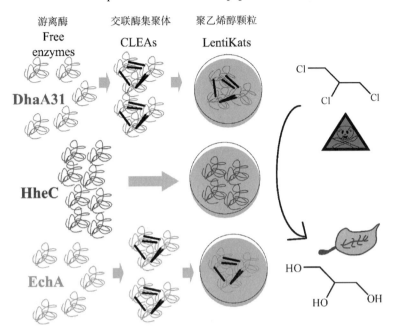

图 4-3-29　初次在体外合成生物降解途径

Fig. 4-3-29　1st *in vitro* synthetic biodegradation pathway assembled

第5章 光谱、色谱及质谱检测技术

Chapter 5　Detection Technology of Spectrum, Chromatography and Mass Spectrometry

本章提要

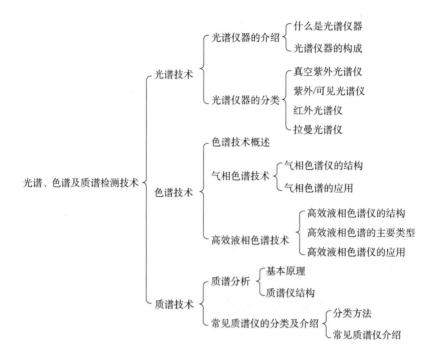

Summary of Chapter 5

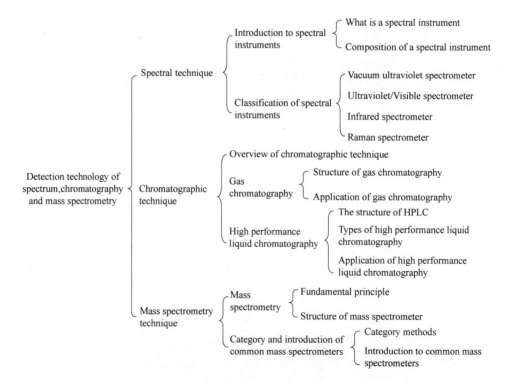

1 光谱技术
1 Spectral Technique

1.1 光谱仪器的介绍
1.1 Introduction to Spectral Instruments

1.1.1 什么是光谱仪器
1.1.1 What Is a Spectral Instrument

光谱仪器是利用光的辐射、色散和吸收等原理，对物质的成分和含量进行定性和定量分析的光学式分析仪器。光谱仪器在冶金、地质、化工、医药、海洋、环境保护、半导体、生物医学、航天探测和同位素应用等许多领域中发挥了重要作用。

Spectrum instrument can conduct the qualitative and quantitative analysis of the composition and content of a substance using the principles of radiation, dispersion and absorption of light. Spectral instruments play an essential role in many fields, such as metallurgy, geology, chemical industry, medicine, ocean, environmental protection, semiconductor, biomedicine, space exploration and isotope application.

1.1.2 光谱仪器的构成
1.1.2 Composition of a Spectral Instrument

光谱仪器的基本作用是测量被研究光（所研究物质反射、吸收、散射或受激发的荧光等）的光谱特性，包括波长、强度等。因此，光谱仪器具有以下功能。①分光：把被研究光按一定波长或波数的分布规律在一定空间内分开。②感光：将光信号转换成易于测量的电信号，测量出相应波长光的强度，得到光能量按波长的分布规律。③绘谱线图：把分开的光波及其强度按波长或波数的分布规律记录保存或显示对应光谱图。

The primary function of the spectral instrument is to measure the spectral characteristics of the studied light (reflected, absorbed, scattered or excited fluorescence, etc.), including wavelength, intensity and so on. Therefore, spectral instruments should have the following functions. ①Light splitting: separate the studied light in a certain space according to the distribution rule of a certain wavelength or wave number. ②Photosensitivity: convert the optical signal into the easily measured electrical signal, measure the intensity of the corresponding wavelength light and obtain the distribution rule of the light energy according to the wavelength. ③Spectral diagram: record or display the corresponding spectral diagram of the separated light waves and their intensity according to the distribution rule of wavelength or wave number.

要具备上述功能，光谱仪器一般由四部分组成：光源和照明系统、分光系统、探测和显示系统，如图 5-1-1 所示。

For the above functions, spectral instruments generally are composed of four parts: light source and lighting system, spectral system, detection and display system, as shown in Fig. 5-1-1.

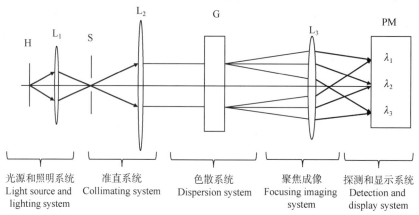

图 5-1-1 光谱仪器的基本结构
Fig. 5-1-1 Basic structure of spectral instrument

光源和照明系统可以是研究的对象，也可以作为研究的工具照射被研究的物质。在研究物质的发射光谱如气体火焰、交/直流电弧以及电火花等激发试样时，光源就是研究的对象；而在研究吸收光谱、拉曼光谱或荧光光谱时，光源则作为照明工具（如汞灯、红外干燥灯、钨灯、氙灯、LED 灯、激光照明灯等）。

Light source and lighting system can be the object of research, or used as research tools to illuminate the studied substance. The light source is the research object when studying the emission spectrum of materials such as gas flame, AC/DC arc and electric spark. In the study of absorption spectrum, Raman spectrum or fluorescence spectrum, the light source is used as lighting tools (such as mercury lamp, infrared drying lamp, tungsten lamp, xenon lamp, LED lamp, laser illuminator, etc.).

分光系统是光谱仪器的核心部分，由准直系统、色散系统、聚焦成像系统三部分组成，主要作用是将照射来的光在一定空间内按照一定波长规律分开。准直系统一般由入射狭缝和准直物镜组成，入射狭缝位于准直物镜的焦平面上。光源和照明系统发出的光通过狭缝照射到准直物镜，变成平行光束投射到色散系统上。色散系统的作用是将入射的单束复合光分解为多束单色光。多束单色光经过成像物镜按照波长的顺序成像在透镜焦平面上；这样，单束复合光经过分光系统后成功变成了多束单色光的像。

Spectral system is the core part of spectral instruments, which consist of collimation system, dispersion system and focusing imaging system. Its main function is to separate the irradiated light in a certain space according to a certain wavelength rule. The collimation system is generally composed of an incident slit and a collimating objective lens. The incident slit is located on the focal plane of the collimating objective lens. The light from the light source and the lighting system shines through the slit to the collimating objective lens and is projected onto the dispersion system as parallel beams. The dispersion system decomposes the incident single composite beams into multiple monochromatic beams. The multiple monochromatic beams are imprinted on the focal plane of the lens according to the wavelength sequence through the imaging objective lens. Therefore, the single composite beams passe through the spectral system and become the image of multiple monochromatic beams.

探测接收系统的作用是将成像系统焦平面上接收的光谱能量转换成易于测量的电信号,并测量出对应光谱组成部分的波长和强度,从而获得被研究物质的特性参数,如物质的组成成分及其含量、物质的温度、星体的运动速度等。

Detection receiving system converts the spectral energy received in the focal plane of the imaging system into easily measurable electrical signals and measures the wavelength and intensity of corresponding spectral components, so as to obtain the characteristic parameters of the studied material, such as the composition and content of the material, the temperature of the material, the velocity of the star and so on.

传输存储显示系统是将探测接收系统测量出来的电信号经过初步处理后存储或通过高速传输接口传至上位机,在上位机上对光谱数据进行进一步处理。

Transmission, storage and display system stores the measured electrical signals from the detection and receiving system after preliminary processing or transfers them to the upper computer through high-speed transmission interface and further processes the spectral data on the upper computer.

1.2 光谱仪器的分类
1.2 Classification of Spectral Instruments

根据所能工作的光谱范围,光谱仪器可分为真空紫外光谱仪、紫外/可见光谱仪、可见光光谱仪、近红外光谱仪、红外光谱仪、远红外光谱仪等。以下重点介绍应用范围较广的几种。

According to the work range of the spectrum, spectral instruments can be divided into vacuum ultraviolet spectrometer, ultraviolet/visible (UV-VIS) spectrometer, visible spectrometer, near infrared spectrometer, infrared spectrometer, far-infrared spectrometer, etc. Sereral applications of widely used ones are introcluced below.

1.2.1 真空紫外光谱仪
1.2.1 Vacuum Ultraviolet Spectrometer

紫外光是由近紫外线、远紫外线和真空紫外线三个波段的光组成的,其波长一般都在 380 nm 以下。如果 UV 的波长在 200 nm 以下,那么在大气中极易被空气吸收,所以该波段的紫外光只有在无氧或真空中才能存在,并进行较长距离的传播。因此,通常把 30~200 nm 这一波段的紫外光称为真空紫外光,或远紫外光。

UV light is composed of near ultraviolet, far ultraviolet and vacuum ultraviolet of three wavelengths of light, usually under 380 nm wavelength, If the UV wavelength is under 200 nm, it is easily absorbed by the air in the atmosphere, so the ultraviolet lights in band can only exist in the anaerobic or vacuum environment and make a long distance transmission, so the UV light in the band of 30-200 nm is often called vacuum ultraviolet light, or far ultraviolet light.

一般的光谱仪是在正常的大气压条件下工作的,而真空紫外光谱仪是在真空室抽真空的条件下工作,满足了真空紫外光传播的条件,可以进行该波段的 UV 和相关光学材料的性能研究。

Generally spectrometers work under normal atmospheric pressure, while vacuum ultraviolet spectrometers work under the vacuum chamber, which meets the conditions of vacuum ultraviolet transmission and can conduct the performance research of UV and related optical materials in this band.

工作原理:真空紫外线在大气中能使空气中的氧气转化为臭氧和原子氧,臭氧能吸收远紫外线生成氧气和活性原子氧,从而使真空紫外线在通过大气层时被吸收。紫外线的波长越短,通过空气的距离越长,被吸收的就越多。根据这一原理,真空紫外光谱仪是设计在真空室内进行的 UV 测试。被测光的波长越短,真空室的真空度要求越高。

Principle: the vacuum ultraviolet ray can make the oxygen in the air into atoms of ozone and oxygen, while ozone absorbs far UV generating oxygen and active oxygen atom, which make the vacuum ultraviolet ray is absorbed by the atmosphere, The shorter the wavelength of ultraviolet ray, the longer the distance through the air, the more are absorbed. According to this principle, the vacuum ultraviolet spectrometer is designed to conduct UV test in a vacuum chamber and the shorter the wavelength of measured light is, the higher the vacuum degree of the vacuum chamber is required.

真空紫外光谱仪(图 5-1-2)一般由氘灯光源、真空紫外光栅单色仪、样品室、光电倍增管及高压电源组成。其中真空紫外光栅单色仪包括入射狭缝、出射狭缝、凹面光栅、光栅扫描驱动机构、主真空室、超高真空阀、扩散泵、真空阀等。

Vacuum ultraviolet spectrometer (Fig. 5-1-2) generally consists of deuterium light source, vacuum ultraviolet grating monochromator, sample chamber, photoelectric shell brightening and high-voltage power supply. Vacuum ultraviolet grating monochromator includes incident slit, exit slit, concave grating, grating scanning drive mechanism, main vacuum chamber, ultra-high vacuum valve, diffusion pump, vacuum valve and so on.

图 5-1-2 真空紫外光谱仪

Fig. 5-1-2 Vacuum ultraviolet spectrometer

1.2.2 紫外/可见光谱仪
1.2.2 Ultraviolet/Visible Spectrometer

紫外/可见光谱仪是指根据物质分子对波长为200~760 nm的电磁波的吸收特性所建立起来的一种定性、定量和结构分析的仪器。紫外可见吸收光谱仪是紫外可见光谱仪中用途较广的一种，其主要由光源、单色器、吸收池、检测器以及数据处理及记录（计算机）等部分组成。

Ultraviolet/Visible spectrometer is an instrument for qualitative, quantitative and structural analysis based on the matter molecules' absorption characteristics of the electromagnetic waves with wavelength of 200~760 nm.

紫外/可见光谱仪（图5-1-3）主要用于化合物的鉴定、纯度检查、异构物的确定、位阻作用的检测、氢键强度的测定以及其他相关的定量分析中，但通常只是一种辅助分析手段，还需借助其他分析方法，如红外、核磁、EPR等对待测物进行分析，以得到精准的数据。

Ultraviolet/Visible (Fig. 5-1-3) spectrometer is mainly used for compound identification, purity examination, isomer detevmination, the detection of steric effect, the determination of hydrogen bonding strength and other relevant quantitative analysis. But it is usually one auxiliary analysis method only, still need to be assisted by other analysis methods, such as infrared, nuclear magnetic, EPR, etc., to get accurate data.

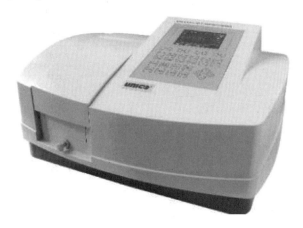

图 5-1-3 紫外/可见光谱仪

Fig. 5-1-3 Ultraviolet/Visible spectrometer

紫外/可见光谱仪广泛用于土壤中各种微量和常量的无机物和有机物质的测定、无机矿物和有机物质的定性和结构分析以及土壤化学过程（络合-解析、溶解沉淀、酸碱离解常数等），也用于植物营养诊断和营养品质分析，如蛋白质、淀粉、可溶性糖、维生素C和铁、锰、铜、锌、硼等元素的分析以及根系活力和多种酶活性的测定。

UV-VIS spectrometer is widely used in detection of all kinds of major and trace elements of the inorganic and organic material in the soil, qualitative and structural analysis of the inorganic mineral and organic material and soil chemical process (complexing-analytical, dissolving precipitation, soda acid dissociation constant, etc.), and is also used in plant nutrition diagnosis and analysis of nutritional quality, such as protein, starch, soluble sugar, vitamin C and iron, manganese, copper, zinc, boron and other elements and determination of root activity and a variety of enzyme activity.

1.2.3 红外光谱仪
1.2.3 Infrared Spectrometer

红外光谱仪是利用物质对不同波长的红外辐射的吸收特性进行分子结构和化学组成分析的仪器。图 5-1-4 所示为红外光谱仪原理。

Infrared spectrometer is an instrument used to analyze the molecular structure and chemical composition of infrared radiation of different wavelengths by using the absorption characteristics of materials. Fig. 5-1-4 shows the principle of infrared spectrometer.

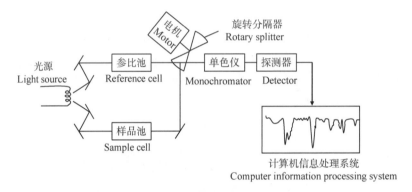

图 5-1-4　红外光谱仪原理

Fig. 5-1-4　Principle of infrared spectrometer

红外光谱仪（图 5-1-5）通常由光源、单色仪、探测器和计算机信息处理系统组成。根据分光装置的不同，其可分为色散型和干涉型。对色散型双光路光学零位平衡红外分光光度计而言，当样品吸收了一定频率的红外辐射后，分子的振动能级发生跃迁，透过的光束中相应频率的光被减弱，造成参比光路与样品光路相应辐射的强度差，从而得到所测样品的红外光谱。

Infrared spectrometer (Fig. 5-1-5) consists of light source, monochromator, detector and computer information processing system. According to the different light-splitting devices, it is divided into types of dispersion and interference. For double light path optical zero-balanced dispersion type infrared spectrophotometer, after the sample absorbs the infrared radiation of a certain frequency, the vibration of the molecular level transits, which makes the frequency of the beam of light abate, causing the relevant radiation intensity difference the reference light path and the sample light path, thus the infrared spectrum of the sample is obtained.

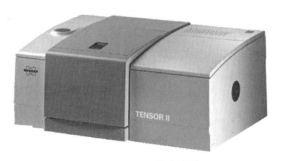

图 5-1-5 红外光谱仪

Fig. 5-1-5 Infrared spectrometer

红外光谱仪具有以下优点：①只需三个分束器即可覆盖从紫外到远红外的区段；②专利干涉仪，连续动态调整功能，且稳定性极高；③可实现 LC/FTIR、TGA/FTIR、GC/FTIR 等技术联用；④智能附件，即插即用、自动识别、仪器参数自动调整；⑤光学台一体化设计，主部件对针定位，无须调整。

Advantages of the infrared spectrometer：①Only three beam splitters can cover the range from ultraviolet to far-infrared；② Patented interferometer, with continuous and dynamic adjustment function, and high stability；③LC/FTIR, TGA/FTIR, GC/FTIR and other technologies；④Smartaccessories with function of plug and play, automatic identification, automatic adjustment of instrument parameters；⑤Integrated design of optical platform, needle positioning of main components without adjustment.

红外光谱仪的应用范围：①进行化合物的鉴定，进行未知化合物的结构分析；②进行化合物的定量分析，进行化学反应动力学、晶变、相变、材料拉伸与结构的瞬变关系研究；③工业流程与大气污染的连续检测；④在煤炭行业对游离二氧化硅的监测；⑤水晶石英羟基的测量，聚合物的成分分析，药物分析等。

Applications of infrared spectrometer：①Identification of compounds and structural analysis of unknown compounds；②Conduct quantitative analysis of compounds, study the transient relationship between chemical reaction kinetics, crystal change, phase change, material tensile and structure；③Continuous detection of industrial process and air pollution；④Monitoring of free silica in the coal industry；⑤Crystal quartz hydroxyl measurement, polymer composition analysis, drug analysis, etc.

1.2.4 拉曼光谱仪

1.2.4 Raman Spectrometer

拉曼光谱是一种散射光谱。拉曼光谱分析法是基于印度科学家 C.V. 拉曼所发现的拉曼散射效应。通过对与入射光频率不同的散射光谱进行分析以得到分子振动、转动方面的信息，并应用于分子结构研究的一种分析方法。最常用的拉曼光谱区域波长是 2.5~25 μm（中红外区）。

Raman spectra is a scattering spectrum. Raman spectrum analysis is based on the Raman scattering effect discovered by Indian scientist C. V. Raman (Raman). By analyzing the scattering spectrum different from the incident light frequency, molecular vibration and rotation information can be obtained and applied to the molecular structure research. The most commonly used wavelength of Raman spectrum is 2.5–25 μm (mid-infrared region).

拉曼散射是分子对光子的一种非弹性散射效应。当用一定频率的激发光照射分子时，一部分散射光的频率和入射光的频率相等。这种散射是分子对光子的一种弹性散射。只有分子和光子间的碰撞为弹性碰撞，没有能量交换时，才会出现这种散射。该散射称为瑞利散射。还有一部分散射光的频率和激发光的频率不等。这种散射称为拉曼散射。

Raman scattering is an inelastic scattering effect of molecules on photons. When the excited light is applied to the molecule at a certain frequency, some of the scattered light has the same frequency as the incident light. This scattering is an elastic scattering of molecules to photons. This scattering occurs only when the collisions between molecules and photons are elastic collisions and there is no energy exchange. This scattering is called Rayleigh scattering. There is also a part of the scattered light whose frequency is not the same with the excitation light frequency. This scattering is called Raman scattering.

拉曼光谱仪（图5-1-6）一般由激光光源、样品装置、滤光器、单色器（或干涉仪）和检测器组成。图5-1-7所示为激光显微拉曼光谱仪光路。

Raman spectrometer (Fig. 5-1-6) generally consists of the laser source, sample device, filter, monochromator (or interferometer) and detector. Fig. 5-1-7 shows optical path of laser microscope Raman spectrometer.

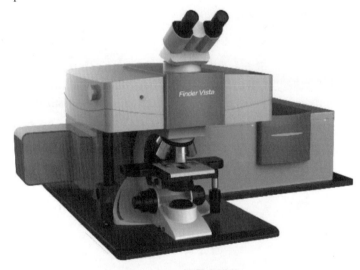

图 5-1-6　拉曼光谱仪
Fig. 5-1-6　Raman spectrometer

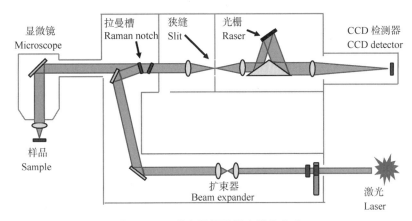

图 5-1-7 激光显微拉曼光谱仪光路

Fig. 5-1-7 Optical path of laser microscope Raman spectrometer

拉曼光谱检测有以下优点：①一些在红外光谱中为弱吸收或者不稳定的谱带（例如：由 C≡N，C=S，S—H 伸缩振动产生的谱带）在拉曼光谱中表现为强谱带，更利于被检出；②拉曼光谱低波数方向的测定范围宽，有利于提供重原子的振动信息；③水的拉曼散射极弱，因此拉曼光谱特别适用于水溶液中样品的研究（水的红外吸收特别强烈）；④与红外光谱相比，对检测物质结构的变化更敏感，分辨率也更强；⑤可直接检测固体样品，无须制样。

Advantages of Roman spectrum detection：①Some spectrum bands（e. g. spectrum bands produced by C≡N, C=S, S—H stretching vibration）are strongly spectrum bands in the Raman spectrum, which is more easyly to be detected；②The measurement range of low wave number direction of Raman spectrum is wide, which is conducive to providing vibration information of heavy atoms；③The Raman scattering of water is very weak, so Raman spectrum is particularly suitable for the study of samples in aqueous solution. (the infrared absorption of water is particularly strong)；④Compared with infrared spectrum, it is more sensitive to changes in the structure of detected substances and has stronger resolution；⑤It can directly test solid samples without sample preparation.

由拉曼光谱可以获得有机化合物的各种结构信息：①同种分子的非极性键所产生的拉曼谱带强度随单键、双键、三键依次增强；②一些在红外光谱中较弱或者不稳定的化学键（e.g. C≡N，C=S，S—H）在拉曼光谱中是强谱带；③环状化合物的对称呼吸振动，一般是最强的拉曼谱带；④醇和烷烃的拉曼光谱结果很相似：C—O 键与 C—C 键的力常数或强度没有很大的差别；羟基和甲基的质量仅差两个单位；与 C—H 和 N—H 的谱带相比，O—H 的拉曼谱带较弱。

Various structural information of organic compounds can be obtained by Raman spectum：①The strength of Raman band generated by non-polar bonds of the same molecule increases with single bond, double bond and triple bond successively；②Some chemical bonds (e. g. , C≡N, C=S, S—H) which are weak or unstable in the infrared spectrum are strongly spectral bands in the Raman spectrum；③The symmetrical respiratory vibration of cyclic compounds is usually the strongest Raman band; ④Raman spectrum results of alcohol and alkane are very similar; there is no great difference between the force constant or strength of C—O and C—C; The hydroxyl and methyl groups differ in mass by only two units; Compared with C—H and N—H, the Raman band of O—H is weak.

拉曼光谱对高分子材料的鉴定如下。

Identification of polymer materials is shown below.

(1) 化学结构和立构性判断：高分子中的 C=C、C—C、S—S、C—S、N—N 等骨架对拉曼光谱非常敏感，常用来研究高分子的化学组分和结构。

(1) Judgment of chemical structure and vertical structure: the skeletons of polymer, such as C=C, C—C, S—S, C—S, and N—N, are very sensitive to Raman spectrum and are commonly used to study the chemical components and structure of polymer.

(2) 组分定量分析：拉曼散射强度与物质浓度线性相关，从而可检测高分子材料的组分含量。

(2) Component quantitative analysis: Raman scattering intensity is linearly correlated with material concentration, so that the component content of polymer materials can be detected.

(3) 晶相与无定形相的表征以及聚合物结晶过程和结晶度的监测。

(3) Characterization of crystal phase and amorphous phase and monitoring of polymer crystallization process and crystallinity.

(4) 动力学过程研究：伴随高分子反应的动力学过程如聚合、裂解、水解和结晶等。相应的拉曼光谱某些特征谱带会有强度的改变。

(4) Kinetic process research: the kinetic processes accompanying the polymer reactions, such as polymerization, pyrolysis, hydrolysis and crystallization. The intensity of some characteristic bands of Raman spectrum will change accordingly.

(5) 聚合物共混物的相容性以及分子相互作用研究。

(5) Study on compatibility and molecular interaction of polymer blends.

拉曼光谱对生物分子的鉴定如下。

Identification of biomolecules is shown below.

(1) 蛋白质二级结构：α-螺旋、β-折叠、无规卷曲及 β-回转。

(1) Secondary structure of protein: α- helix, β- folding, random curl and β-rotation.

(2) 蛋白质主链构象：酰胺Ⅰ、Ⅲ、C—C、C—N 伸缩振动。

(2) The main chain of the protein structure: amide Ⅰ, Ⅲ, C—C and C—N stretching vibration.

(3) 蛋白质侧链构象：苯丙氨酸、酪氨酸、色氨酸的侧链和后两者的构象及存在形式随其微环境的变化而变化。

(3) Protein side-chain structure: the structure and existence form of the side chains of phenylalanine, tyrosine and tryptophan change with their microenvironment.

(4) 对构象变化敏感的羧基、巯基、S—S、C—S 的构象变化。

(4) Changes of carboxyl group, sulfhydryl group, S—S and C—S structures who are sensitive to structural changes.

（5）生物膜的脂肪酸碳氢链旋转异构现象。

(5) Rotation isomerism of fatty acid hydrocarbon chain in biofilm.

（6）DNA 分子结构以及和 DNA 与其他分子间的作用。

(6) Molecular structure of DNA and its interactions with other molecules.

（7）研究脂类和生物膜的相互作用、结构、组分等。

(7) Study the interaction, structure and composition of lipids and biofilms.

（8）对生物膜中蛋白质与脂质相互作用提供相关信息。

(8) Provide relevant information on protein and lipid interactions in biofilms.

Ⅱ 色谱技术
Ⅱ Chromatographic Technique

2.1 色谱技术概述
2.1 Overview of Chromatographic Technique

生物产品的分离、纯化是现代生物技术中一种常见的技术。色谱技术在各种分离分析技术中是应用最为广泛的技术手段，适用于多组分混合样品的分离分析，具有分离效率高、灵敏度高、分析速度快、操作简便等优点，被广泛应用于生物样本的分析检测中，如核苷、核苷酸、氨基酸的分离分析、蛋白质的分离纯化、生化分析等。

The separation and purification of biological products is the most common technique in modern biotechnology. Chromatography is the most widely used technique in various kinds of separation analysis, which is suitable for the separation analysis of multi-component mixed samples. It has the advantages of high separation efficiency, high sensitivity, rapid analysis speed and simple operation and is widely used in the analysis and detection of biological samples, such as nucleoside, nucleotide, amino acid separation analysis, protein separation and purification, biochemical analysis.

色谱的分离过程是基于样品中的各个组分与互不相溶的两相间相互作用的差异而实现的。随着两相间的相对移动，各组分在两相间反复多次分配，同时分配差异不断累积，最终使各组分在固定相上分离。图 5-2-1 所示为色谱分离示意图。

The process of chromatographic separation is based on the differential interactions of the components with the two phases that are not mutually soluble. With the relative movement of the two phases, the components are distributed repeatedly on the two phases and the distribution difference accumulates continuously. Finally, the components are separated in the stationary phase. Fig. 5-2-1 shows schematic of chromatographic separation.

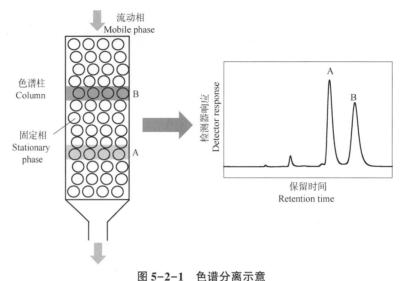

图 5-2-1　色谱分离示意

Fig. 5-2-1　Schematic of chromatographic separation

混合物中各组分经色谱分离后，随流动相依次流出，即得到色谱流出曲线（图 5-2-2）。色谱图是色谱技术对化合物进行定性定量分析的基础。同一物质在相同色谱条件下，色谱峰的保留值是一定的，因此色谱峰的保留值可以作为物质定性的依据；同时，物质的含量是和色谱峰的峰高或峰面积线性相关的，由此可以进行定量分析。

After separation, the components in the mixture are flushed with the mobile phase and the chromatogram (Fig. 5-2-2) is obtained. Chromatogram is the basis of qualitative and quantitative analysis of compounds in chromatography technique. The retention value of chromatographic peak of the same chemical is constant under the same chromatographic condition, which can be used as the qualitative basis for different chemicals. While the concentration of chemicals is linearly related to the peak height or peak area, which can be used in the quantitative analysis.

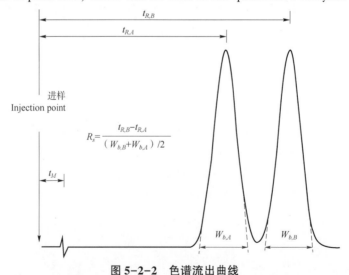

图 5-2-2　色谱流出曲线

Fig. 5-2-2　Schematic of a general chromatogram

按流动相的不同可以将色谱分为气相色谱、液相色谱及超临界流体色谱。本节我们介绍其中应用最为广泛的色谱技术——气相色谱（GC）和高效液相色谱（HPLC）。

According to the different mobile phase, chromatography can be divided into gas chromatography, liquid chromatography and supercritical fluid chromatography. In this section, we introduce gas chromatography (GC) and high performance liquid chromatography (HPLC), which are the most widely used chromatographic techniques.

2.2 气相色谱技术
2.2 Gas Chromatography

流动相为气体的色谱法称为气相色谱，该法适用于分析气体及易挥发有机物。

GC is a kind of chromatography in which the mobile phase is gas. This method is suitable for analysis of gases and volatile organic chemicals.

2.2.1 气相色谱仪的结构
2.2.1 Structure of Gas Chromatography

气相色谱仪一般由载气系统、进样系统、分离系统、温控系统、检测系统及数据处理系统组成，如图 5-2-3 所示。

Gas chromatography typically consists of carrier gas system, sample injection system, separation system, temperature control system, detection system and etc, as shown in Fig. 5-2-3.

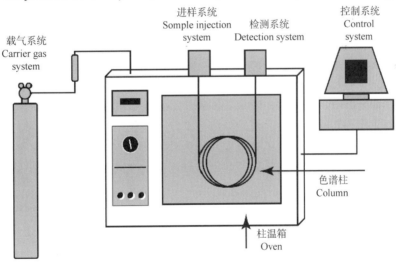

图 5-2-3　气相色谱仪

Fig. 5-2-3　Gas chromatography

载气系统包括气源、气体净化器以及气路控制系统。载气作为气相色谱分离中的流动相，需要选择与检测器匹配，且不干扰分析物检测的惰性气体；同时，载气的纯度、流速对色谱柱的分离性能、检测器的灵敏度也有很大影响，气路控制系统的作用就是将载气及辅助气进行稳压、稳流及净化，以满足气相色谱分析的要求。

Carrier gas system includes gas source, gas purifier and gas flow control system. As the mobile phase in gas chromatographic separation, carrier gas should to be an inert gas that matches the detector and does not interfere with the detection of analytes. Moreover, the purity and flow rate of carrier gas also have a great impact on the separation performance of column and the sensitivity of detector. Good flow control system is needed in GC to provide a constant or well-defined flow of carrier gas and auxiliary gas to meet the requirements of gas chromatography.

进样系统包括进样装置与汽化室，可使样品注射进入后立刻汽化，并被载气带入色谱柱。在毛细管气相色谱仪中，毛细管柱样品容量小，为避免过载最常采用的进样方式是分流式进样，样品在汽化室内汽化后大部分经分流管道放空，一小部分进入色谱柱。

The injection system consists of an injection device and a gasification chamber. The sample is vaporized immediately after injection and carried into the column. In capillary gas chromatograph, the capacity of column is low. In order to avoid overload, the most commonly used injection method is split injunction, most of the sample in the gasification room is vented through the split pipe and a small portion enters the column.

色谱柱是色谱仪的核心，其作用是分离样品。气相色谱所用的色谱柱主要包括填充柱和毛细管柱。填充柱是里面装填有固定相的色谱柱，柱容量大、分离效果好；毛细管柱是固定液均匀地涂在毛细管内壁的空心柱，与填充柱相比，其渗透性好、传质阻力小，但柱容量低，对检测器的灵敏度要求比较高。

The column is the core of the chromatograph and its function is to separate samples. The columns used in GC are mainly include packed columns and capillary columns. Packed columns are packed with stationary phases, with large capacity and good separation performance. Capillary columns are open-tube columns. Its consist of a column that has the stationary phase coated on its interior surface. Compared with the packed columns, capillary columns have good permeability and low mass transfer resistance, but its column capacity is low, thus higher sensitivity of detector is required.

在气相色谱分析中，温度是一个重要的操作参数，温度会影响样品的汽化过程、气相色谱柱的分离及检测器的灵敏度。温控系统主要是对汽化室、色谱柱炉及检测器进行温度控制。其中，色谱柱温的控制方式有恒温和程序升温两种方式，对宽沸程的样品可采用程序升温，即在一个分析周期内柱温随时间由低到高呈线性或非线性增加，达到用最短时间获得最佳分离的目的。

Temperature is a significant operating parameter in chromatography. Temperature will affect the gasification process of sample, the separation on chromatography column and the sensitivity of detector. The temperature control system is mainly for the gasification room, column furnace and detector temperature control. The temperature of column is controlled by constant temperature and temperature programming. Temperature programming usually used to separate a mixture of chemicals with a wide range of boiling points, that is, the column temperature increases linearly or nonlinearly with time in one analysis cycle, so as to achieve the optimal separation in less time.

样品经色谱柱分离后，各组分依次随载气进入检测器，其保留时间以及浓度或质量信息被转换为电信号传递给计算机，进一步处理后得到样品的色谱流出曲线。对气相色谱检测器的性能要求为灵敏度高、检出限低、响应线性范围宽、响应速度快、稳定性好、普适性好。根据检测原理的不同可分为浓度型检测器与质量型检测器。浓度型检测器的响应值与组分浓度成正相关,如热导检测器和电子捕获检测器;质量型检测器的响应值和单位时间内进入检测器的组分的量正相关,如氢火焰离子化检测器和火焰光度检测器。

Each component enters the detector with carrier gas after separation and its retention time, concentration or mass information are converted into electrical signals and transmitted to the computer. Finally, the chromatographic outflow curve of the sample is obtained. The requirements of gas chromatography detector are high sensitivity, low detection limit, wide response linear range, fast response, good stability and good universality. According to their detection principles, gas chromatography detectors can be classified into two types: concentration detectors and mass detectors. The response of concentration detector is positively correlated with the component concentration, such as thermal conductivity detector and electron capture detector. While the response value of mass detectors is positively correlated with the amount of components entering the detector per unit time, such as hydrogen flame ionization detector and flame photometric detector.

2.2.2 气相色谱的应用
2.2.2 Application of Gas Chromatography

A. 食品分析

A. Food analysis

气相色谱用于食品分析的范围很广。食品中重要的营养成分如氨基酸、脂肪酸、糖类都可以用气相色谱进行分析。此外,很多食品添加剂也可以用气相色谱检测。

Gas chromatography is used in a wide range of food analysis. Important nutrients in foods such as amino acids, fatty acids and saccharides can be analyzed by gas chromatography. In addition, many food additives can also be detected by gas chromatography.

B. 环境样品分析

B. Environmental analysis

气相色谱法适用于环境中挥发性污染物的检测,如苯系污染物、挥发性有机酸、家装材料中的一些挥发性有机化合物等。

Gas chromatography is suitable for the detection of volatile pollutants in the environment, such as benzene pollutants, volatile organic acids, some volatile organic compounds in home decoration materials, etc.

C. 生物、医学样品分析

C. Biological and medical analysis

气相色谱法在生物、医学中应用也很广泛,如用于分离和测定生物体中的氨基酸、维生素、糖类等化合物。

Gas chromatography is also widely used in biology and medicine, such as for the separation and determination of amino acids, vitamins, saccharides and other chemicals in organisms.

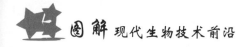

D. 石油化工样品分析

D. Petrochemical analysis

在石油化工行业中，气相色谱法可用于气体、馏分油及原油的检测，也可用于烃类、非烃类等组分的分析，具有良好的分离效果。

In the petrochemical industry, gas chromatography can be used for the detection of gas, distillate and crude oil, but also for the analysis of hydrocarbon, non-hydrocarbon and other components, with good separation performance.

2.3 高效液相色谱技术
2.3 High Performance Liquid Chromatography

液相色谱是指流动相是液体的色谱法。高效液相色谱技术是在经典液相色谱的基础上，引入气体色谱的理论和技术，实现液相色谱的高效化。该方法适合分析气相色谱难以分析的挥发性差、热稳定性差、具有生物活性的物质，填补了气相色谱的空缺，迅速成为目前应用最为广泛的色谱技术。

Liquid chromatography is a kind of chromatographic method in which the mobile phase is liquid. High performance liquid chromatography is based on classical liquid chromatography, the theory and technology of gas chromatography is introduced to achieve high performance. This method is suitable for the analysis of compounds with poor volatile, thermal instability and biological activity, that fills the vacancy of gas chromatography and quickly becomes the most widely used chromatography technology.

2.3.1 高效液相色谱仪的结构
2.3.1 The Structure of HPLC

高效液相色谱仪由高压输液系统、进样系统、分离系统、检测系统及数据处理系统组成，如图5-2-4所示。

HPLC consists of high pressure mobile phase delivery system, sample injection system, separation system, detection system and data processing system, as shown in Fig. 5-2-4.

高压输液系统由流动相储存器、高压输液泵、梯度洗脱装置及压力表组成。其作用是在高压输液泵的驱动下，把流动相连续送入液路系统中。液相色谱仪所用的色谱柱内径较细，且其中装填的固定相粒度较小，因此，需要采用高压输液泵使流动相快速稳定地流过固定相。

The high pressure mobile phase delivery system consists of mobile phase reservoir, high pressure delivery pump, gradient elution device and pressure gauge. The function of the delivery system is to continuously deliver the mobile phase into the liquid system by high pressure delivery pump. The column used in liquid chromatography has a small inner diameter and the stationary phase size is small. So it is necessary to use the high pressure infusion pump to make the mobile phase flow through the stationary phase quickly and stably.

进样系统包括进样口、注射器和进样阀，高效液相色谱中一般采用六通阀进样，进样后样品随流动相进入色谱柱，在色谱柱上进行分离。

The injection system includes inlet, syringe and injection valve. Generally, six-way valve is used in HPLC to inject samples. After the sample is injected, the sample enters and passes through the column with the mobile phase.

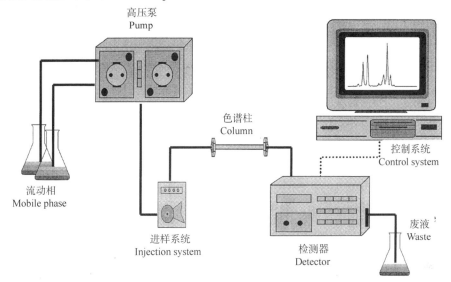

图 5-2-4　高效液相色谱仪

Fig. 5-2-4　**High performance liquid chromatography**

分离系统包括色谱柱、恒温器及连接管等。色谱柱是色谱分离的核心部件，柱子通常是直型的不锈钢管，内部装填有不同类型的固定相。根据分析样品的不同可选择不同类型的固定相来满足分析检测需求。

The separation system includes column, thermostat and connection tube etc. Column is the core of the chromatographic separation, which is usually consist of a straight stainless steel tube packed with different types of stationary phase inside. Different types of stationary phase can be selected to meet the needs of analysis and detection requirements according to the different chemicals.

样品中各组分依先后次序进入检测器，进入检测器的信号被记录下来，得到液相色谱图，流出液流入废液瓶。理想的检测器应具有灵敏度高、重复性好、响应快、线性范围宽等特点。用于高效液相色谱的检测器主要分为两种类型。一种类型是溶质性质检测器，它对被分离组分的物理或化学特性有响应，如紫外、荧光、电化学检测器等；另一种类型是总体检测器，是对样品和流动相总体的物理或化学特性有响应，如示差折光检测器、电导检测器等。

Each component is separated on the chromatographic column and entered the detector. The signals are recorded and the liquid chromatogram is obtained. The outflow liquid flowed into the waste liquid bottle. The ideal detector should have high sensitivity, good repeatability, fast response and wide linear range. There are two main types of detectors for HPLC. The first type of detectors are responsive to the physical or chemical properties of the separated components, such as ultraviolet, fluorescence, electrochemical detectors, etc. The other type of detectors are

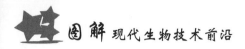

responsive to the physical or chemical properties of the sample and the mobile phase, such as differential refractive index detector, conductivity detector, etc.

2.3.2 高效液相色谱的主要类型
2.3.2 Types of High Performance Liquid Chromatography

在高效液相色谱分析中，如何选择最佳的固定相、色谱条件及分离模式是实现理想分离的关键。常用的液相色谱分离模式主要有如下几个。

In HPLC analysis, how to choose the best stationary phase, chromatographic condition and separation mode is the key to realize ideal separation. The commonly used separation modes are as follows.

A. 吸附色谱

A. Adsorption chromatography

该类色谱以吸附剂为固定相，样品组分在吸附剂表面发生吸附和脱附过程，一般用于极性不同的物质的分离。

Adsorption chromatography is a kind of liquid chromatography in which the stationary phase is adsorbent. The components of a sample are retained based on their adsorption and desorption on the surface of the adsorbent. This mode is generally used for the separation of samples with different polarity.

B. 分配色谱

B. Partition chromatography

该类色谱的固定相由载体和固定相组成，固定液或涂覆或键合于载体表面。样品组分的分离基于溶质在流动相和固定相之间的分配差异，可用于多种类型样品的分离分析，是目前应用最广泛的色谱类型。

The stationary phase of partition chromatography consists of carrier and liquid stationary phase, which is coated or bonded onto the surface of the carrier. The separation of each component is based on their partitioning between liquid mobile phase and stationary phase. This mode can be used for the separation of many types of samples, which is the most widely used chromatographic type at present.

C. 离子交换色谱

C. Ion-exchange chromatography

该类色谱的固定相表面带有带电荷基团及游离的平衡离子。样品中的带电离子可与游离的平衡离子进行可逆交换，因此可根据不同离子与离子交换剂间亲和力的差别将其分离。

The surface of the stationary phase of ion-exchange chromatography contains fixed charge groups and free counter-ions. Ions in the sample can be reversibly exchanged with free counter-ions, so different ions can be separated according to the different affinity with fixed charge groups.

D. 空间排阻色谱

D. Size-exclusion chromatography

该类色谱的固定相为多孔凝胶，一般用于大分子化合物的分离。样品中体积大的组分无法进入凝胶孔道中，在色谱柱中流动的路径短、保留时间短、体积小的组分可以进入孔道，保留时间长，从而达到分离的目的。

The stationary phase of size-exclusion chromatography is a porous silica gel, which is commonly used for the separation of macromolecular compounds. The components with large size in the sample cannot enter the pores of gel, so the flow path in the column is short and the retention time is shorter. While the components with small size can enter the pore and have longer retention time, thus the purpose of separation is achieved.

E. 亲和色谱

E. Affinity chromatography

亲和色谱的固定相载体上连接有能专一性识别目标物质的配基，因此能够从生物样品中分离和分析目标物质，可用于生物活性物质的分离和测定。图 5-2-5 所示为亲和色谱分离模式。

The stationary phase of affinity chromatography is bonded with ligands that can identify target components specifically, so that the target components can be separated and analyzed from biological samples and can be used for separation and detection of the biologically active compounds. Fig. 5-2-5 shows the separation modes of liquid chromatography.

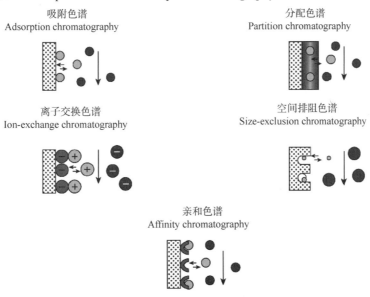

图 5-2-5　亲和色谱分离模式

Fig. 5-2-5　Separation modes of liquid chromatography

2.3.3　高效液相色谱仪的应用
2.3.3　Application of High Performance Liquid Chromatography

高效液相色谱的应用范围较气相色谱的应用更广泛，主要有如下几个。

The application range of HPLC is wider than that of GC. The main applications are as follows.

A. 食品分析

A. Food analysis

高效液相色谱可用于食品中关键营养成分的分析，如蛋白质、糖类、氨基酸、微生物、有机酸、有机胺等；也可用于食品添加剂的分析，如防腐剂、甜味剂、色素、抗氧化剂等；此外，还可用于食品污染物的分析，如霉菌毒素、多环芳烃、微量元素等。

HPLC can be used for the analysis of key nutrients in foods, such as proteins, saccharides, amino acids, microorganisms, organic acids, organic amines, etc. It can also be used for the analysis of food additives such as preservatives, sweeteners, pigments, antioxidants, etc. Moreover, it also be used for the analysis of food contaminants, such as mycotoxins, polycyclic aromatic hydrocarbons, trace elements, etc.

B. 环境分析

B. Environment analysis

高效液相色谱在环境分析中起到非常重要的作用，如用于分析多环芳烃及农药残留。

HPLC plays a very important role in environmental analysis, for example in the analysis of polycyclic aromatic hydrocarbons and pesticide residues.

C. 生命科学

C. Life science

高效液相色谱在生命科学研究中既可用于小分子的分析检测，如氨基酸、糖类、维生素等，也可用于大分子检测，如蛋白质、脂质、核酸等。

In the life science research, HPLC can be used for the analysis and detection of small molecules, such as amino acids, saccharides, vitamins, etc. It can also be used for macromolecule detection, such as proteins, lipids, nucleic acids, etc.

Ⅲ 质谱技术
Ⅲ Mass Spectrometry Technique

3.1 质谱分析
3.1 Mass Spectrometry

3.1.1 基本原理
3.1.1 Fundamental Principle

质谱分析是一种通过测定分子的质荷比（m/z）来进行物质定性、定量分析的技术手段，其既可以测得分子的质量信息，又可以利用串联质谱使分子发生碎裂，通过测定碎片粒子的质量而获得分子的结构信息。

Mass spectrometry (MS) is a qualitative and quantitative analytical method by measuring mass over charge (m/z) of molecules. Not only molecular mass could be obtained by MS, but also its structure could be determined by analyzing masses of fragment ions in tandem MS.

质谱分析的基本过程如下。离子源将待测样品分子离子化，产生的带电离子在离子传输装置的导引下进入质量分析器。在质量分析器中，不同质荷比的离子在电场或磁场作用下由于具有不同的运动特性而发生分离，并随即进入离子检测器形成对应的电信号，该信号通过电路系统的放大及处理，最终得到样品的分析结果，并以谱图的形式呈现，即质谱图（图5-2-6）。在质谱图中，横坐标为测得离子的质荷比（m/z），纵坐标代表离子的相对信号丰度，反映了同一质谱图中各离子的相对丰度（或含量）的高低。

The main process of MS analysis is described as below. The analytes firstly are ionized by ionization source and the generated charged ions are then transferred into mass analyzer by ion transfer devices. In mass analyzer, ions with different m/z are separated and analyzed according to their different motion properties in electric or magnetic field. The analyzed ions finally go into ion detector and produce corresponding electrical signals, which are then further processed and amplified by electronics system and presented in the mass spectrum as the final analytical results (Fig. 5-2-6). In a mass spectrum, x-axis represents m/z of ions and y-axis represents the relative intensity of detected ion signals, which reveals the relative abundances (or concentrations) of different ions in the same spectrum.

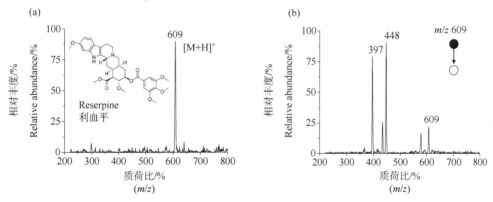

图 5-2-6　质谱图

Fig. 5-2-6　Mass spectrum

(a) 典型质谱图；(b) 串联质谱图

(a) The typical mass spectrum；(b) Tandem mass spectrum of reserpine

3.1.2 质谱仪结构
3.1.2 Structure of Mass Spectrometer

质谱仪的种类很多，但基本结构相同，如图5-2-7所示。质谱仪主要由离子源、离子传输透镜、质量分析器、检测器、真空系统以及控制与数据处理系统等部分构成。由于大量的中性分子会影响离子在电场及磁场中的运动特性，因此质谱仪的许多部件需要处于真空环境中，主要包括离子传输装置、质量分析器和检测器。事实上，早期的离子源（如电子轰击离子源、化学离子源等）也需要工作于真空条件下，但后来随着离子化技术的发展,各种敞开式离子源不断出现并应用于质谱仪。敞开式离子源在大气压条件下即可完成样品的离子化,从而摆脱了对真空条件的依赖,极大地拓展了质谱仪的应用范围。

Mass spectrometers can be classified to many different types, as shown in Fig. 5-2-7, but they have the same structure, as shown in Fig. 5-2-7. A mass spectrometer is typically composed of ion source, ion optics, mass analyzer, ion detector, vacuum system, control and data system, etc. Because natural molecules would affect the ion motions in electric and magnetic fields, most components of a mass spectrometer, mainly including ion transfer devices, mass analyzer and ion detector, must be operated under the vacuum condition. In fact, the earliest ion sources, for instance electron impact ionization (EI) and chemical ionization (CI), are also need to work in vacuum. With the development of different techniques for ionization, various ambient ionization sources come into being and are applied into mass spectrometers. Getting rid of dependence on vacuum, ambient ionization sources enable the ionization process at atmospheric environment and thus greatly broaden the application areas of mass spectrometers.

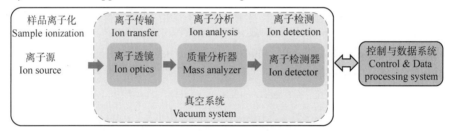

图 5-2-7　质谱仪结构

Fig. 5-2-7　Structure of the mass spectrometer

在质谱仪诸多结构部件中,离子源与质量分析器的作用最为关键,二者共同决定了质谱仪的应用范围及关键分析性能,下面将对二者作主要介绍。

Of most components of a mass spectrometer, ionization source and mass analyzer are two most crucial ones determining the application range and analytical performances of a mass spectrometer. Therefore, the details of these two components are discussed below.

A. 离子源

A. Ionization source

离子源是完成样品由中性分子向离子转化的核心部件。不同的离子源采用不同的离子化原理,从而直接决定了质谱仪的样品分析类别与应用范围。根据离子化能量的大小,离子源可分为高能离子源(硬电离)和低能离子源(软电离),二者的典型代表分别为电子轰击离子源和化学离子源。

Ionization source is the core component converting neutral analyte molecules to charged ions. Based on different principles, different ionization sources determine the sample category analyzed by a mass spectrometer and thus its application range. Based on ionization energy, ionization source could be classified as high energy ionization (also called hard ionization) source and low energy ionization (soft ionization) source. EI and CI are two typical ones of these two types of ionization sources, respectively.

早期的质谱仪主要采用以 EI 为主的真空高能离子源,只能离子化气体及挥发性有机物等非极性小分子,无法分析极性生物大分子,如蛋白质、核酸等。电喷雾离子化(ESI)与基质辅助激光解析离子化(MALDI)两项技术的出现,打破了质谱用于生物大分子分析的瓶颈,促进了基于质谱分析的生物组学的兴起。两项技术的发明者,美国科学家 John B. Fenn 与日本科学家田中耕一也因此共同获得了 2002 年的诺贝尔化学奖。ESI 和 MALDI 是目前生物学领域应用最广泛的离子化技术,其中 ESI 可以产生多电荷的分子离子峰,从而降低了超大分子检测对于质谱仪质量范围的苛刻要求,可以实现超大质量分子(如超大蛋白质)的分析,因此已成为蛋白质组学研究的首选质谱技术。图 5-2-8 所示为利用 ESI 源测得的超大蛋白聚合体 GDH 和 GroEL 的质谱图。

The ionization sources employed in earlier mass spectrometers were high energy ones majored as EI, thus only nonpolar and small molecules of gas and volatile organics, could be ionized. Large biological molecules, such as protein and nucleic acids, could not be analyzed by these ionization sources. Above problem was not solved until the developments of electrospray ionization (ESI) and matrix assisted laser desorption ionization (MALDI), which broke the bottleneck of mass spectrometric analysis for biological large molecules and thus boosted the emergence of biological omics. The inventors of the two techniques, scientists of John B. Fenn from the U.S. and Koichi Tanaka from Japan, were awarded Nobel Prize in Chemistry of 2002. ESI and MALDI are the most widely used ionization methods in biological fields, of which ESI can produce multiple charged molccular ion peak and decrease the strict demand of detecting macromolecules on mass range of a mass spectrometer. ESI enables the detection of macromolecules, for example macro proteins and has become the 1st choice for MS analysis in proteomics research. Fig. 5-2-8 shows the mass spectra of macro proteins of GDH and GroEL by ESI.

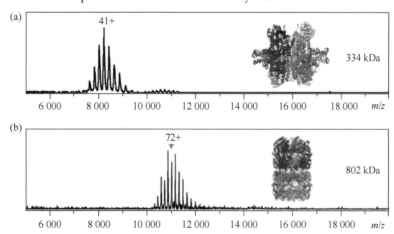

图 5-2-8 利用 ESI 源测得的超大蛋白聚合体 GDH 和 GroEL 的质谱图
Fig. 5-2-8 The mass spectra of macro proteins of GDH and GroEL by ESI
(a) GDH; (b) GroEL
(a) GDH; (b) GroEL

针对不同的应用需求，目前广泛应用的离子源还有大气压化学离子源（APCI）、纳升电喷雾离子源（nano-ESI）、电感耦合等离子体离子源（ICP）、解析电喷雾离子源（DESI）、实时直接分析离子源（DART）等。

For different application needs, some other ionization sources are also widely used, including atmospheric pressure chemical ionization (APCI), nano electrospray ionization (nano-ESI), inductively coupled plasma (ICP), desorption electrospray ionization (DESI) and direct analysis in real time (DART), etc.

B. 质量分析器

B. Mass analyzers

质量分析器是质谱仪最核心的部分，主要完成分子的质量分析过程，其工作原理及结构精度等因素会影响质谱仪的整体分析性能。表5-2-1所示为常见质量分析器性能参数比较。根据工作原理的不同，质量分析器主要类型有扇形磁偏转、四极杆、离子阱、飞行时间、轨道阱、傅里叶变换离子回旋共振。在一些高性能质谱仪中，往往会将多种不同的质量分析器串联使用，以达到特定的分析功能。例如，由四极杆和飞行时间串联而成的四极杆-飞行时间（Q-TOF）质量分析器、由三个四极杆串联而成的三重四极杆（QQQ）质量分析器等。由于工作原理不同，各种质量分析器的性能不尽相同。评价质量分析器性能的指标参数主要包括质量范围、质量分辨率、质量精确度、离子采样模式、工作气压及串联质谱功能等。

A mass analyzer is the most key element in a mass spectrometer, whose function is to complete the process of mass analysis for ions. The working principle and structural accuracy of mass analyzer would greatly influence the whole instrumental performances. Table 5-2-1 shows comparison of analytical performance of common mass analyzers. According to the difference in their working principles, mass analyzers could be categorized as: magnetic sector, quadrupole, ion trap, time of flight (TOF), orbitrap and Fourier transform ion cyclotron resonance (FTICR). For some high-performance mass spectrometers, two or more analyzers are usually used together to achieve certain analytical functions. For example, QQQ is consisted of three quadrupole analyzers and Q-TOF is a combination of TOF analyzer serially connected with a quadrupole. Determined by different working modes, different kinds of mass analyzers have different analytical performances, which can be evaluated by a series of parameters, including mass range, mass resolution, mass accuracy, ion sampling, working pressure and tandem MS.

表 5-2-1 常见质量分析器性能参数比较
Fig. 5-2-1 Comparison of analytical performances of common mass analyzers

项目	扇形磁场 Magnetic sector	四极杆 Quadrupole	离子阱 Ion trap	飞行时间 TOF	轨道阱 Orbitrap	傅里叶变换离子回旋共振 FTICR
质量范围 Mass range	20 000	4 000	6 000	>100 000	50 000	30 000
质量分辨率 Mass resolution	100 000	2 000	4 000	10 000	100 000	500 000
质量精确度 Mass accuracy/ppm①	<10	100	100	5~50	< 5	< 5
离子采样 Ion sampling	连续 Continuous	连续 Continuous	脉冲 Pulsed	脉冲 Pulsed	脉冲 Pulsed	脉冲 Pulsed
工作气压 Pressure/Torr②	10^{-6}	10^{-5}	10^{-3}	10^{-6}	10^{-10}	10^{-10}
串联质谱 Tandem MS	MS/MS	MS/MS	MS^n	MS/MS	MS/MS	MS/MS

3.2 常见质谱仪的分类及介绍
3.2 Category and Introduction of Common Mass Spectrometers

3.2.1 分类方法
3.2.1 Category Methods

A. 按分析原理

A. Analytical principle based classification

质量分析器的类别决定了质谱仪的分析原理，因此基于分析原理分类的质谱仪通常以质量分析器而命名，基本命名规则为"质量分析器名+质谱仪"。基于该分类方法，目前常见的质谱仪主要有磁偏转质谱仪、四极杆质谱仪、离子阱质谱仪、飞行时间质谱仪、傅里叶变换离子回旋共振质谱仪以及轨道阱质谱仪。除了采用单一质量分析器外，一些现代质谱仪结构更为复杂，往往会集成多个质量分析器，这类仪器称为杂化质谱仪。目前的杂化质谱仪主要采用多个原理相同或不同的质量分析器在空间上串联排列而成，以实现不同的分析功能，其基本分类命名规则为"质量分析器1-质量分析器2-……+质谱仪"，如"四极杆-飞行时间质谱仪""离子阱-轨道阱质谱仪"等。图 5-2-9 所示为 Thermo Fisher

---

① 1 ppm = 10^{-6}。

② 1 Torr ≈ 133.3 Pa。

公司生产并销售的一款线性离子阱-轨道阱杂化质谱仪的结构图，其采用了一个线性离子阱和一个高分辨轨道阱作为质量分析器，同时集成了多个离子传输部件，包括四极杆滤质器、传输多极杆等。

The type of mass analyzer determines the analytical principle of the mass spectrometer. Therefore, mass spectrometers classification based on analytical principles are usually named after the mass analyzer and the basic naming convention is "Mass Analyzer Name +Mass Spectrometer". Based on the classification method, the current common mass spectrometers are mainly magnetic sector mass spectrometer, quadrupole mass spectrometer, ion trap mass spectrometer, time-of-flight mass spectrometer, Fourier transform ion cyclotron resonance mass spectrometer and orbitrap mass spectrometer. In addition to the use of single mass analyzer, some modern mass spectrometers are more complex in structure and often integrate multiple mass analyzers. These instruments are called hybrid mass spectrometers. To realize different analytical functions, the current hybrid mass spectrometer often uses a plurality of mass analyzers with the same principle or different. These analyzers are spatially and serially arranged in order. The basic classification naming rule of hybrid mass spectrometer is "Mass Analyzer 1-Mass Analyzer 2…+Mass Spectrometer", such as "quadrupole-TOF mass spectrometer" and "ion trap-orbitrap mass spectrometer". Fig. 5-2-9 shows the structural schematic of a linear ion trap-orbitrap hybrid mass spectrometer produced and sold by Thermo Fisher, which uses a linear ion trap and a high-resolution orbitrap as mass analyzers. In addition, this instrument integrates multiple ion transport components, including quadrupole mass filters, multipole ion guides, etc.

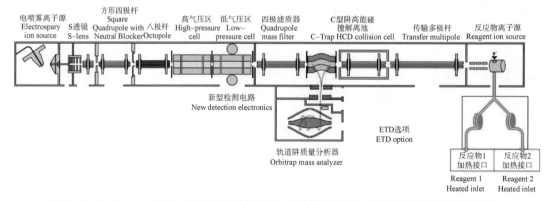

图 5-2-9　Thermo Fisher 公司生产并销售的一款线性离子阱-轨道阱杂化质谱仪结构图

Fig. 5-2-9　The structural schematic of a linear ion trap-orbitrap hybrid mass spectrometer produced and sold by Thermo Fisher

B. 应用领域分类

B. Application based classification

根据应用领域的不同，质谱仪主要可分为有机质谱仪、无机质谱仪以及生物质谱仪三类。

According to their application fields, mass spectrometers could be classified into three types：

organic mass spectrometer, inorganic mass spectrometer and biological mass spectrometer.

有机质谱仪主要用于有机物分析,是各类质谱仪中数量最多、用途最广的一类。目前有机质谱仪广泛应用于食品安全、药物代谢、环境科学以及临床医学等领域。在实际应用中,有机质谱通常会与其他各种技术连用分析,因此根据其应用特点的不同又可分为气相色谱-质谱联用仪、液相色谱-质谱联用仪、基质辅助激光解析离子化质谱仪等。无机质谱仪主要用于无机物分子及元素分析,主要类型有电感耦合等离子体质谱仪(ICP-MS)和二次离子质谱仪(SIMS)。

Organic mass spectrometers are mainly used for organic matter analysis and are the most widely used one of all types of mass spectrometers. By far, organic mass spectrometers are widely used in food safety, drug metabolism, environmental science and clinical medicine. In practical applications, organic mass spectrometers are usually coupled with a variety of other techniques. Based on their application characteristics, organic mass spectrometers can be divided into gas chromatography-mass spectrometer (GC-MS), liquid chromatography-mass spectrometer (LC-MS), matrix-assisted laser desorption ionization mass spectrometer (MALDI-MS), etc. Inorganic mass spectrometers are mainly used for the analysis of inorganic molecules and elements. The main types are inductively coupled plasma mass spectrometer (ICP-MS) and secondary ion mass spectrometer (SIMS).

ICP-MS 被广泛应用于各领域的微量元素分析,具体原理及仪器结构将在下文详细介绍。二次离子质谱仪通过连续或脉冲的一次离子束轰击分析物表面产生二次离子,并对二次离子进行分析,其具有微量成分检测、高表面灵敏度、同位素检测以及空间分子分布信息检测等功能和优点,因此被应用于金属、盐类、有机化合物、制药、聚合物、电子材料、催化剂以及生化组织样品成像等分析领域。

ICP-MS is widely used in the analysis of trace elements in various fields. The specific principles and instrument structure will be described in detail below. The secondary ion mass spectrometer bombards the analyte surface by continuous or pulsed primary ion beam to generate secondary ions and analyzes the secondary ions. It has advantages of trace component detection, high surface sensitivity, isotope detection and spatial molecular distribution information detection. Therefore, SIMS has been applied to the analyses of metals, salts, organic compounds, pharmaceuticals, polymers, electronic materials, catalysts and biochemical tissue sample imaging.

生物质谱仪是用于精确测量生物大分子,如蛋白质、核酸、糖类等的分子量,并提供分子结构信息的一类质谱仪。由于生物样品的分子量比较大,常规离子源无法对其离子化,因此生物质谱仪相对于其他类型质谱仪的发展较晚。根据离子化方法,目前应用成熟的生物质谱仪主要有电喷雾离子化质谱仪和基质辅助激光解析离子化质谱仪两大类。

Biological mass spectrometer is a type of mass spectrometer used to accurately measure the molecular mass of biological macromolecules, such as proteins, nucleic acids, sugars, etc., and also to provide molecular structural information. Due to the relatively large molecular mass of biological samples, conventional ion sources cannot ionize them, so biological mass spectrometers have evolved later than other types. Based on the ionization method, the mature biological mass spectrometers currently used are two major types: electrospray ionization mass spectrometer (ESI-

MS) and matrix-assisted laser desorption ionization mass spectrometer.

3.2.2 常见质谱仪介绍
3.2.2 Introduction to Common Mass Spectrometers

A. 气相色谱-质谱仪
A. Gas chromatography-mass spectrometer

气相色谱-质谱仪是将气相色谱仪与质谱仪联用，以同时实现样品的色谱分离与质谱高灵敏检测的仪器装置。气相色谱-质谱技术是现在样品分离与鉴定的主流方法之一，主要应用于环境分析、农药检测、植物代谢物分析、有机酸检测等领域。

Gas chromatography-mass spectrometer is the device that combines a gas chromatograph with a mass spectrometer to simultaneously perform chromatographic separation of samples and highly sensitive MS detection. GC-MS is one of the mainstream methods for sample separation and identification mainly used in environmental analysis, pesticide testing, plant metabolite analysis, organic acid detection, etc.

如图 5-2-10（a）所示，常规气相色谱-质谱仪主要由气相色谱分离部分和质谱仪检测器两部分连用组成，其中气相色谱主要包含进样接口、色谱柱、管柱烘箱等部分组成。在分析过程中，挥发性样品或气态样品由进样接口注入，并通过加热快速汽化，随后进入气相色谱柱，分离后的不同分析物最终通过气相色谱与质谱的传输管线进入质谱仪而被检测。气相色谱-质谱仪的离子源通常为电子轰击离子源与化学离子源，其中电子轰击离子源应用最为广泛。根据质量分析器的不同，目前的气相色谱-质谱仪主要有四极杆、飞行时间、轨道阱等类型，其中气相色谱-四极杆质谱仪发展最为成熟，应用也最为广泛。图 5-2-10（b）为安捷伦公司推出的一款气相色谱-质谱仪（型号：7890A），其采用了电子轰击离子源与四极杆质量分析器。

As shown in Fig. 5-2-10 (a), the conventional gas chromatography-mass spectrometer is mainly composed of a gas chromatographic separation part and a mass spectrometer detector. The gas chromatograph mainly includes a sample inlet, a column, column oven and other components. During the analysis, volatile or gaseous samples are injected from the injection interface and rapidly vaporized by heating and then enter the GC column. The separated analytes are finally passed through the transfer line of the gas chromatograph and enter into mass spectrometer for detection. The ionization sources used in GC-MS are usually EI and CI, of which EI is the most widely used one. According to different mass analyzers, current gas chromatography-mass spectrometers mainly include types of quadrupoles, time-of-flight and orbitrap, etc., among which gas chromatography-quadrupole mass spectrometers are the most mature and widely used. Fig. 5-2-10 (b) shows a gas chromatography-mass spectrometer (model 7890A) from Agilent that uses an electron impact ion source and a quadrupole mass analyzer.

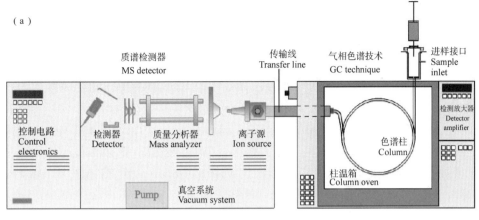

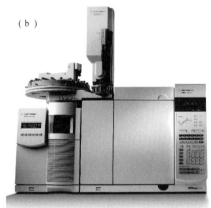

图 5-2-10 气相色谱-质谱仪

Fig. 5-2-10 Conventional gas chromatography-mass spectrometer

(a) 气相色谱-质谱仪原理结构图；(b) 安捷伦 7890A 气相色谱-四极杆质谱仪

(a) The schematic of GC-MS；(b) The 7890A GC-MS of Agilent

B. 电喷雾离子化质谱仪

B. Electrospray ionization mass spectrometers

电喷雾离子化质谱仪因采用电喷雾离子源而得名，通常与液相色谱连用，用于液态样品的分离及检测，广泛应用于蛋白质组学、代谢组学等领域，是目前最主流的质谱仪器之一。

Electrospray ionization mass spectrometer (ESI-MS) is named after electrospray ionization source it used. It is usually coupled with liquid chromatography for separation and detection of liquid samples. ESI-MS is one of the most mainstream mass spectrometers and has been widely applied in fields of proteomics, metabolomics, etc.

不同于 GC-MS 通常所采用的真空离子源，ESI-MS 采用了 ESI 这一大气压离子源，从而可以实现样品的常压离子化，避免了液体直接进样分析对于质谱仪真空系统的损害。在 ESI-MS 中，常压条件产生的离子需要通过特殊的接口传输至仪器真空腔体中的质量分析器，该接口称为大气压接口。大气压接口不仅是维持仪器真空环境的必备组成，而且是决定离子传输效率和检测灵敏度的关键因素,是质谱仪的重要部件之一。目前商业化的ESI-MS仪器普遍采用了连续大气压接口,该接口基于多级(通常为3~4级)差动真空腔室,主要由进样毛细管(或锥孔)、锥孔透镜、离子传输装置等部分组成。图5-2-11所示为电喷雾离子阱质谱原理,在ESI源处离子化后的样品离子由不锈钢毛细管进入一级真空腔体,并通过锥孔采样透镜进入二级真空腔内,最后在八极杆及透镜组的聚焦导引下进入三级腔体内的离子阱质量分析器和检测器,完成最终分析。

Different from the vacuum ionization source commonly used in GC-MS, ESI-MS uses ESI, an atmospheric pressure ionization source, realizing sample ionization at atmosphere and avoiding damage of direct liquid injection to the vacuum of mass spectrometer. In ESI-MS, the ions generated under ambient conditions need to be transmitted to the mass analyzer in the vacuum chamber of the instrument through a special interface. This interface is called atmospheric pressure interface (API). API is one of the most important components for the mass spectrometer, which is not only the essential component to maintain instrumental vacuum, but also a key factor determining ion transmission efficiency and detection sensitivity. Currently commercialized ESI-MS instruments widely integrate the continuous atmospheric pressure interface based on a multi-stage (usually 3-4) differential vacuum chamber and mainly consist of sampling capillary (or skimmer), skimmer lens, ion transfer devices and other parts. Fig. 5-2-11 shows the structure of an electrospray ion trap mass spectrometer. The sample ions ionized at the ESI source enter the 2nd stage vacuum chamber by a stainless steel capillary and enter the 2nd stage vacuum chamber through the skimmer lens. Finally, the ions are focused and guided by octopole and other lenses into ion trap mass analyzer and detector in the three-stage vacuum for analysis and detection.

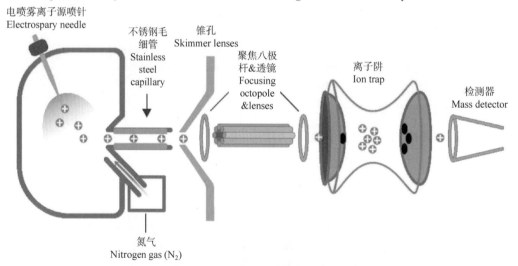

图5-2-11 电喷雾离子阱质谱仪原理

Fig. 5-2-11 The structure of electrospray trap mass spectrometery

C. 基质辅助激光解析离子化-飞行时间质谱仪
C. Matrix-assisted laser desorption ionization-time-of-flight mass spectrometer

MALDI-TOF 是一类用于非挥发性样品，尤其是固体样品分析的重要仪器，其具有高灵敏度、高准确度、高分辨率等特点，广泛应用于蛋白质组学和生物组织成像。MALDI-TOF 质谱仪主要由 MALDI 离子源与 TOF 质量分析器组成，二者均需工作于真空环境。

MALDI-TOF is one kind of important instruments for the analysis of non-volatile samples, especially solid samples. It features high sensitivity, high accuracy and high resolution and thus has been widely used in proteomics and biological tissue imaging. The MALDI-TOF mass spectrometer is primarily composed of a MALDI ionization source and a TOF mass analyzer, both of which need operating in vacuum.

MALDI 离子源的作用是使样品离子化，其基本原理是用激光照射样品与基质形成的共结晶薄膜，基质从激光中吸收能量并传递给样品分子，同时发生激发态基质分子与样品分子之间的电荷转移反应，从而使样品分子离子化。基质和激光的参数是影响 MALDI 离子化效率的主要因素。大部分基质为有机酸，现今最常用的基质有 2,5-二羟基苯甲酸（DHB）、α-氰基-4-羟基肉桂酸（CHCA）、3,5-二甲氧基-4-羟基肉桂酸（SA）、2,4,6-三羟基苯乙酮（THAP）。目前 MALDI 所使用的激光主要是脉冲宽度为 3~5 ns 的紫外激光，最常用的两种波长为 337 nm 和 355 nm。

The function of MALDI ionization source is to ionize analytical samples. In the ionization, a laser is used to illuminate the crystal film formed by the analyte and matrix. The matrix absorbs energy from the laser and transmits it to the sample molecules, while the excited matrix radical molecules react with the sample molecules in the form of charge transfer reaction, finally ionizing sample molecules. Matrix and laser parameters are significant factors affecting the ionization efficiency of MALDI. Most of the matrix is organic acid and the most commonly used ones nowadays are 2,5-dihydroxybenzoic acid (DHB), α-cyano-4-hydroxycinnamic acid (CHCA), 3,5-dimethoxy-4-Hydroxycinnamic acid (SA), 2,4,6-trihydroxyacetophenone (THAP). At present, the laser used for MALDI is mainly ultraviolet laser with a pulse width of 3-5 ns and the two most commonly used wavelengths are 337 nm and 355 nm.

在 MALDI 离子源处离子化后的样品离子，在离子传输装置的引导下进入 TOF 质量分析器及检测器，完成离子的质量分析和最终信号检测。TOF 质量分析器的基本原理是经电场加速后的离子进入离子漂移管。由于不同质荷比的离子到达检测器的飞行时间不同，从而完成不同样品的质量分析。TOF 分析器根据原理及结构的不同主要有线性式、反射式、正交式三种。此外，各种类型 TOF 串联形成的质量分析器阵列也在现代诸多仪器中得到应用，以获得更高的分析性能。图 5-2-12 所示仪器即采用了两级线性 TOF 与一级反射式 TOF 的组合。

With the guidance of ion transport devices, the analyte ions generated at the MALDI ionization source enter the TOF mass analyzer and detector for mass analysis and final signal detection. The typical principle of the TOF mass analyzer is that the ions accelerated by the electric field enter the ion drift tube. Ions with different mass-to-charge ratios reach detector with different flight time, thus full fill mass analysis of the sample. According to their principle and structure, TOF analyzers mainly have three types: linear, reflective and orthogonal ones. In addition, mass analyzer arrays of various types of TOF series are also used in many modern instruments to achieve higher analytical performance. For example, the instrument shown in Fig. 5-2-12 uses a combination of two stage linear TOF and one stage reflective TOF.

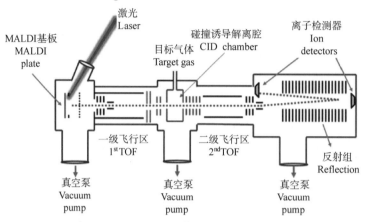

图 5-2-12　MALDI-TOF 质谱仪原理结构

Fig. 5-2-12　The structural schematic of MALDI-TOF

D. 电感耦合等离子体质谱仪

D. Inductively coupled plasma mass spectrometer

电感耦合等离子体质谱仪是用于元素分析的一类极其重要的质谱仪器，因其采用了电感耦合等离子体（ICP）离子源而得名。一般而言，ICP-MS 仪器的组成主要包括采样系统、ICP 离子源、离子传输透镜、质量分析器、检测器以及真空系统等部分。在分析过程中，待测样品通过采样系统进入 ICP 离子源产生的等离子体区，先后经过汽化、去溶剂化、离子化等过程形成带电离子，再由离子透镜等装置传输至质量分析器中进行分析，并最终被检测器检测。根据不同的分析需求，ICP-MS 可采用不同类型的质量分析器，包括四极杆、飞行时间、轨道阱以及傅里叶变换离子回旋共振等。图 5-2-13（a）所示为安捷伦公司的 7800 型 ICP-MS 仪器结构图，该仪器采用了高精度双曲面四极杆作为分析器，可实现元素的定性定量分析。

ICP-MS is an extremely important mass spectrometer for elemental analysis, which is named after the ionization source of inductively coupled plasma it uses. In general, an ICP-MS instrument is mainly composed of sampling system, ICP ionization source, ion tranmission lenses, mass analyzer, detector and vacuum system. In analysis process, the sample to be tested enters the plasma zone generated by the ICP ionization source through the sampling system and forms charged ions after processes of gasification, desolvation, ionization, etc., and the ions are then transferred by ion optics to the mass analyzer for mass analysis, followed by the final detection by detector. Depending on different analytical needs, ICP-MS can employ different types of mass analyzers, including quadrupole, time-of-flight, orbitrap, Fourier transform ion cyclotron resonance, etc. Fig. 5-2-13 (a) shows the structure of the Model 7800 ICP-MS instrument from Agilent, which uses a high-precision hyperbolic quadrupole as mass analyzer for qualitative and quantitative analysis of elements.

ICP 离子源是元素及样品离子化的核心部件。如图 5-2-13（b）所示，常见的 ICP 离子源主要由三层同轴石英管组成，分别用于样品引入、等离子气体引入和气流冷却。石英管最外层缠绕有射频感应线圈，通过射频发生器产生具有足够能量强度的感应磁场，使通入等离子体管中的辅助气体（通常为氩气）发生解离而产生含有高密度电子的离子化气体，在气流作用下形成喷射状等离子体火炬，其温度可达 6 000~10 000 K。产生的等离子体可使进入其中的样品分子获得能量而发生解离，形成离子。

The ICP ionization source is the core component for ionization of elements and samples. As shown in Fig 5-2-13 (b), the ICP ionization source is typically composed of coaxial quartz tubes with three layers, which are used as sample inlet, plasma tube and coolant tube, respectively. The outermost layer of the quartz tube is wound with a RF induction coil, together with RF generator where induced magnetic field having a sufficient energy intensity can be generated. This magnetic field is able to dissociate the auxiliary gas (usually argon) that is introduced into the plasma tube and produce ionized gas containing high-density electrons, forming a jet-like plasma torch under the assistance of the gas stream whose temperature can reach 6,000−10,000 K. The generated plasma can cause the sample molecules entering in it to gain energy and dissociate to form ions.

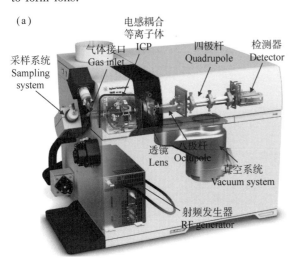
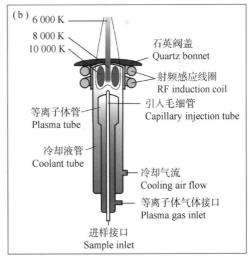

图 5-2-13　安捷伦公司 7800 型 ICP-MS 仪器结构

Fig. 5-2-13　Structure of the model 7800 ICP-MS instrument from Agilent

(a) 安捷伦 7800 型 ICP-MS 结构；(b) ICP 离子源原理结构

(a) The 7800 ICP-MS of Agilent；(b) The schematic of ICP ionization source

第6章 肿瘤治疗分析技术

Chapter 6 Treatment and Analysis Technology of Tumors

本章提要

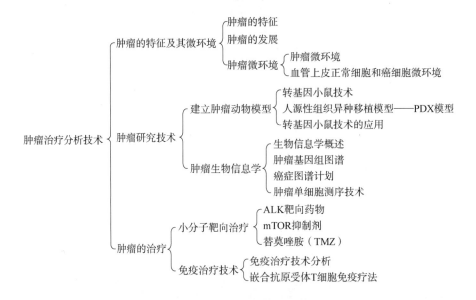

Summary of Chapter 6

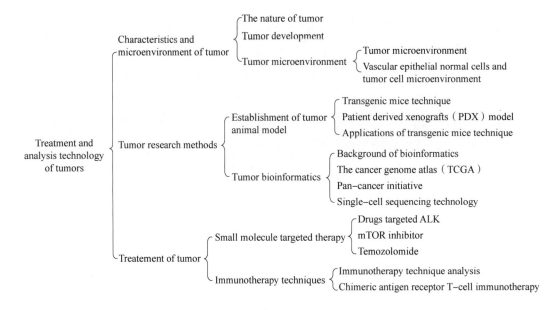

肿瘤的特征及其微环境
Characteristics and Microenvironment of Tumor

1.1 肿瘤的特征
1.1 The Nature of Tumor

相比正常细胞，肿瘤细胞具有许多异常的改变，如图 6-1-1 所示。例如，肿瘤细胞在形状上通常比其起源的细胞大；肿瘤细胞的细胞核也大于正常细胞核；染色体发生重排；细胞聚集时没有明显的边界。由于不会发生接触抑制，肿瘤细胞获得了无限生长和分裂的能力。肿瘤细胞的主要特征包括基因突变、维持增殖信号传导、实现永生化、阻碍程序性凋亡细胞死亡、侵袭性、破坏基因组稳定以增加表型变异等。肿瘤发展的每个阶段都有特征性的分子，遗传和细胞变化相关联，这些变化导致了越来越恶性的表型获得。从一个阶段到另一个阶段的进展是由自发遗传变异和选择进化驱动的。异常细胞特征性的分子不仅仅是一个靶标用于癌症的精确诊断，还对精准治疗起到了指导性的作用。

Compared with normal cells, tumor cells show numbers of abnormal characteristics, as shown in Fig. 6-1-1. For example, the shape and the nucleus of tumor cells are usually larger than the original cells, the chromosomes are rearranged, and cell aggregation blurs the boundary. Since there's no contact inhibition effect, tumor cells obtain the abilities of infinite growth and division. The main characteristics of tumor cells include gene mutation, sustaining growth signaling pathway, immortalization, inhibition of apoptosis, invasiveness, destabilization of the genome to increase phenotypic variation, etc. There are specific molecular, genetic and cellular changes in each stage of tumor development. These changes are associated with increasingly malignant phenotypic acquisition. Progress from one stage to another is driven by spontaneous genetic variation and selective evolution. The molecular characteristic of abnormal cells is not only a target for accurate diagnosis of cancer, but also plays a guidance role in precise medicine.

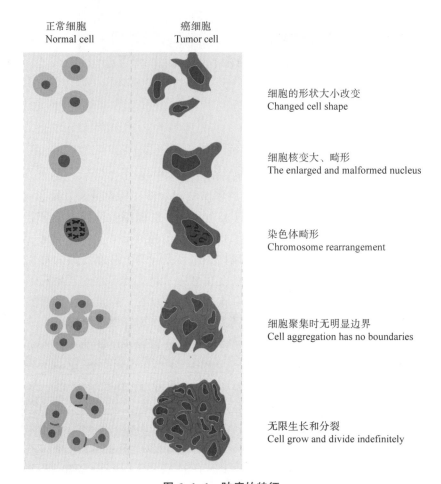

图 6-1-1 肿瘤的特征
Fig. 6-1-1 The Nature of Tumor

1.2 肿瘤的发展
1.2 Tumor Development

癌变的过程通常分为如下四个阶段：第一阶段是癌前阶段，这个阶段的细胞仍然是正常的，只是具备了发生癌变的客观条件；第二阶段是原位癌阶段，细胞开始发生癌变并且具备了癌细胞的特征；第三阶段是早期癌症阶段，癌细胞浸润性生长破坏正常组织结构；第四阶段是转移阶段，癌细胞通过血液或者淋巴结的方式向远处扩散，在原发灶之外的其他器官建立转移灶。图 6-1-2 所示为肿瘤的形态。

The process of carcinogenesis is usually divided into the following four stages: the first stage is precancerous stage, in which the cells are still normal but have the objective conditions for carcinogenesis; the second is the in situ cancer stage, the cells possess the neoplastic characteristics; in the third stage, invasive growth of cancer cells destroys the normal structure of tissues; in the fourth stage, the cancer cells begin to metastasis and damage other organs. Fig. 6-1-2 shows tumor cell shape.

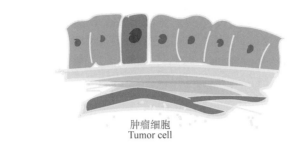
肿瘤细胞
Tumor cell

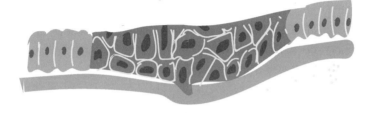
早期肿瘤
Early-stage tumor

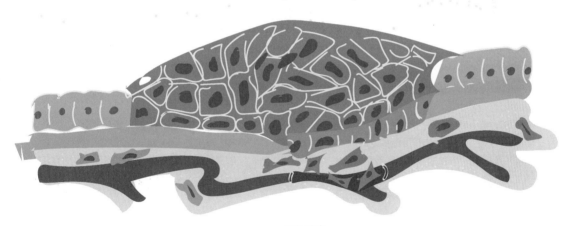
恶性肿瘤
Malignant tumor

图 6-1-2 肿瘤的形态

Fig. 6-1-2 Tumor cell shape

1.3 肿瘤微环境
1.3 Tumor Microenvironment

1.3.1 肿瘤微环境
1.3.1 Tumor Microenvironment

肿瘤微环境（图 6-1-3），又称为肿瘤基质，是促进肿瘤细胞生长与转移的重要支撑因素，同时也是导致肿瘤治疗过程中耐药的主要因素。肿瘤微环境主要由肿瘤基质细胞和非细胞两大部分组成，其细胞成分包括纤维细胞、免疫细胞、内皮细胞等；非细胞主要是细胞外基质和分泌到细胞外的细胞因子，如 CCR5、CXCL12/SDF-1 等。肿瘤微环境的特征是低氧、低 pH 值、肿瘤间质高压、血管高渗透压和炎症反应性等。微环境对肿瘤的发展起到促进作用。近年来研究发现肿瘤微环境是肿瘤转移和耐药的"帮凶"，可针对肿瘤微环境进行肿瘤治疗。

  Tumor microenvironment (Fig. 6-1-3), also known as tumor stroma, is a pivotal support to promote the growth and metastasis of tumor cells, and also a major factor to drug resistance during anti-tumor treatment. The tumor microenvironment is composed of tumor stromal cells and non cellular companents and microenvironmental cells include fibroblasts, immune cells, endothelial cells, etc. Non-cellular components of tumor microenvironment mainly include the extra cellular stroma and cytokines secreted into the extra cellular environment, such as CCR5、CXCL12/SDF-1. Tumor microenvironment is characterized by hypoxia, low pH value, tumor interstitial pressure, and high osmotic pressure of blood vessel. As microenvironmental play an important role in the development of tumor, Recent studies have found that tumor microenvironment is the "accomplice" of tumor metastasis and drug resistance, so tumor microenvironment can be a target for anti-tumor treatment.

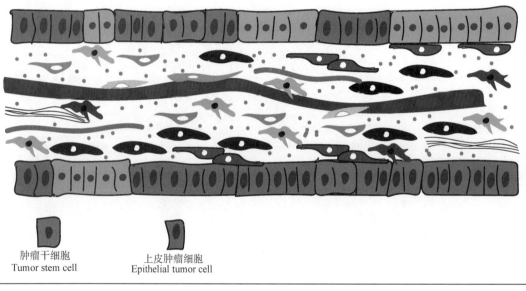

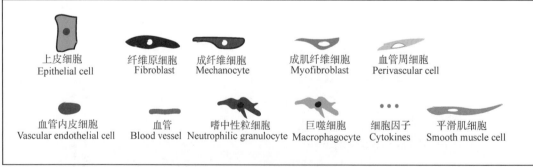

图 6-1-3 肿瘤微环境

Fig. 6-1-3 Tumor microenvironment

1.3.2 血管上皮正常细胞和癌细胞微环境
1.3.2 Vascular Epithelial Normal Cells and Tumor Cell Microenvironment

图 6-1-4 所示为血管上皮正常细胞和癌细胞微环境。正常血管上皮细胞产生的促血管生成因子（Ang-2）在细胞浆中被 Weibel-Palade 小体包裹不会被分泌出去。另外，附着在血管上皮细胞的周细胞分泌的 Ang-1 与 Tie-2 二聚体受体结合，维持了稳定的静息状态。相反，在肿瘤的微环境中，Ang-2 脱离 Weibel-Palade 小体，并与分泌出上皮细胞与上皮细胞膜上的 Tie-1/Tie-2 异源二聚体受体结合，抑制了静息状态，并使周细胞脱离上皮细胞。此外，肿瘤通常处在缺氧的条件下，血管内皮生长因子（VEGF）、肿瘤坏死因子（TNF）能够被诱导分泌，启动了肿瘤微环境中的血管生成，为肿瘤的增殖转移提供了条件。

Fig. 6-1-4 shows vascular epithelial normal cells and tumor cell microenvironment. The angiogenic factor (Ang-2) produced by normal vascular epithelial cells is not secreted but wrapped by the Weibel-Palade body in the cytoplasm. Additionally, the Ang-1 secreted by pericyte, which is attached to the perivascular epithelial cells, binds to the Tie-2 dimer receptor in order to maintain a stable quiescent state. On the contrary, in the tumor microenvironment, Ang-2 is detached from Weibel-Palade bodies, followed by secretion from epithelial cells, and then binds to Tie-1/Tie-2 heterodimer receptors on epithelial cell membrane. It leads to inhibition of quiescent state and detachment of pericytes from epithelial cells. Under hypoxia condition, tumor growth related factors, such as vascular epithelial growth factor (VEGF), tumor necrosis factor (TNF), were secreted to promote angiogenesis, consequently enhance proliferation and metastasis of tumor.

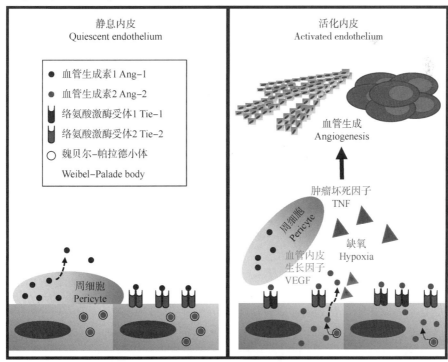

图 6-1-4 血管上皮正常细胞和癌细胞微环境

Fig. 6-1-4 Vascular epithelial normal cells and tumor cell microenvironment

在图 6-1-5（a）所示的正常血液细胞微环境中，血管上皮细胞和成骨细胞通过 Notch 信号通路介导造血干细胞良性增殖。在成骨细胞分泌的趋化因子 SDF-1/CXCL12 介导造血干细胞迁徙归巢。在一系列的细胞因子的共同作用下，造血干细胞维持不分化的状态。

在图 6-1-5（b）所示的肿瘤微环境中，间充质干细胞与肿瘤细胞通过 Notch 信号通路相互作用，介导了肿瘤干细胞的增殖与成骨细胞功能的抑制；血管上皮细胞分泌血管内皮生长因子（VEGF）、成纤维细胞生长因子（FGF），分别作用于肿瘤细胞膜上的相关受体，介导血管生成，促进肿瘤的转移与生长；同时通过 ICAM1 与 LFA1 的作用，肿瘤细胞的恶性增殖得到了启动。

In normal microenvironment shown in Fig. 6-1-5（a）, vascular epithelial cells and osteoblasts mediate the proliferation of normal hematopoietic stem cells through Notch signaling pathway. The chemokine SDF-1/CXCL12, secreted by the osteoblasts, mediates the migration and homing of hematopoietic stem cells. Hematopoietic stem cells won't differentiate under the effect of a series of cytokines.

In tumor microenvironment shown in Fig. 6-1-5（b）, mesenchymal stem cells interact with tumor cells through Notch signaling pathway, which mediates the proliferation of tumor stem cells and the inhibition of osteoblasts. Vascular epithelial cells secrete vascular endothelial growth factor（VEGF）, fibroblasts growth factor（FGF）, which bind to receptors located on the tumor cell membrane, mediates angiogenesis and promotes tumor metastasis and growth. The malignant proliferation of tumor cells is initiated through ICAM1 and LFA1.

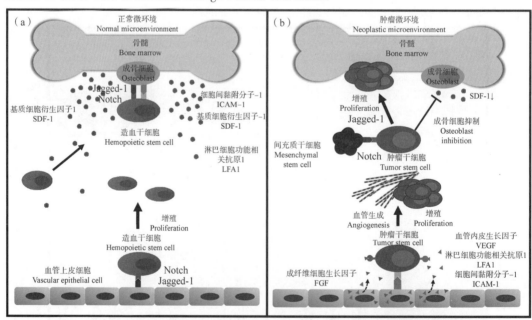

图 6-1-5　造血干细胞骨髓微环境和肿瘤干细胞骨髓微环境

Fig. 6-1-5　Bone marrow microenvironment of hematopoietic stem cells and tumor stem cell

（a）正常血液细胞微环境；（b）肿瘤微环境

（a）Normal microenvironment；（b）Tumor microenvironment

Ⅱ 肿瘤研究方法
Ⅱ Tumor Research Methods

2.1 建立肿瘤动物模型
2.1 Establishment of Tumor Animal Model

2.1.1 转基因小鼠技术
2.1.1 Transgenic Mice Technique

A. 显微注射法

A. Microinjection

图 6-2-1 所示为显微镜注射法建立转基因小鼠的示意。该方法利用显微操作仪将外源基因注射到受精卵中，待其发育成桑葚胚或囊胚后移入假孕母鼠的输卵管，受精卵运行到母鼠子宫，发育成熟致分娩，即可获得转基因小鼠。此方法同样可以与第三代基因编辑技术——CRISPR/Cas9 联用。用预先设计的基因编辑体系替代外源基因，将其注射进入受精卵中，再用同样的方法培养、移植受精卵即可获得基因修改目的的转基因小鼠。

Fig. 6-2-1 shows the establishment of transgenic mice by microinjection. In this method, the heterologous gene is injected into the nucleus of the oosperm by micromanipulator, and then the oosperm will develop into morula or blastula. Transplant the morula or blastula into the fallopian tube of the pseudopregnant mouse and the transgenic mice could be obtained. Microinjection can also be combined with the 3rd generation gene editing technology—CRISPR/Cas9. The transgenic mice with modified target gene could be obtained by transplanting the oosperm after injection of pre-designed gene editing agent.

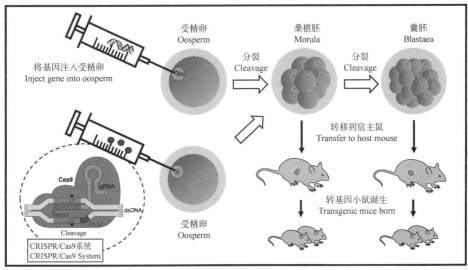

图 6-2-1 显微注射法建立转基因小鼠

Fig. 6-2-1 Establishment of transgenic mice by microinjection

B. 胚胎干细胞介导法

B. Embryonic stem cell mediating method

与受精卵显微注射不同，胚胎干细胞介导法（图6-2-2）注射使用的是在胚胎中获得的胚胎干细胞。该技术的主要流程：在体外将外源基因导入胚胎干细胞，然后将带有外源基因的胚胎干细胞注射到动物囊胚中；携带外源基因的胚胎干细胞参与整个胚胎的发育而形成嵌合体。未完全分化的胚胎干细胞具有全能性。当把胚胎干细胞注入囊胚后，这些干细胞可以分化为各种嵌合体器官（包括生殖腺）。因此，将外源基因导入胚胎干细胞，就可以获得生殖系统整合了外源基因的嵌合体小鼠。通过后续的杂交即可获得具有纯合目的基因的转基因小鼠。

Unlike microinjection of oosperm, embryonic stem cell mediating method (Fig. 6-2-2) uses the embryonic stem cell obtained in an embryo. The main procedures of this technique: *in vitro* transfection of foreign genes into embryonic stem cells, and then injection of transfected embryonic stem cells with foreign genes into blastocysts; embryonic stem cells with exogenous genes participate in the development of the entire blastocyst to form chimeras. Embryonic stem cells are multi-potential. When embryonic stem cells are injected into blastocyst cavity, these stem cells can differentiate into various chimeric organs (including gonads). Therefore, the chimeric mouse with the exogenous gene can be obtained. Subsequently, transgenic mice with homozygous target gene are established after hybridization.

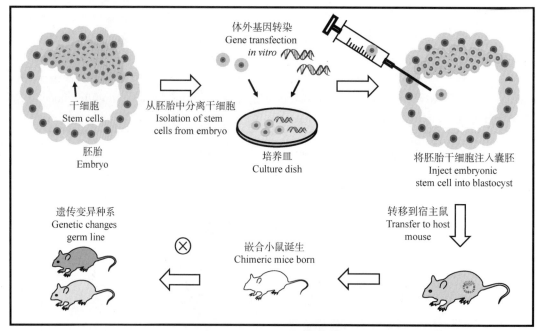

图 6-2-2 胚胎干细胞介导法

Fig. 6-2-2 Embryonic stem cell mediating method

2.1.2 人源性组织异种移植模型
2.1.2 Patient-derived Xenografts Model

人源性组织异种移植模型（PDX 模型），如图 6-2-3 所示，包含的要点：①肿瘤组织来源于病人；②建立模型所用的小鼠是重症免疫缺陷型小鼠（NSG 小鼠）；③肿瘤组织直接以组织的形式移植进入 NSG 小鼠的体内。PDX 模型具有许多优势，如肿瘤直接取自临床病人，不经体外培养直接移植到小鼠体内，从而保持了肿瘤的异质性等。由于肿瘤的临床特征得到极大的保留，因此 PDX 模型更适用于研究肿瘤的生物学特征。

The main characteristics of patient-derived xenografts model (PDX model), as shown in Fig. 6-2-3: ① the tumor tissue originats from the patient; ② the mice used to establish the model are severe immunodeficency mice (NSG mice); ③ The tumor tissue is transferred directly into the body of NSG mice. PDX model has many advantages, such as tumor taken directly from clinical patients and directly transplanted into mice without culture in vitro, thus maintaining the heterogeneity of tumor. Because the clinical features of tumors are greatly preserved, PDX model is more suitable for studying the biological characteristics of tumors.

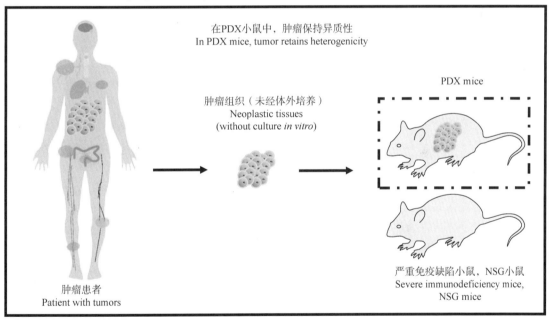

图 6-2-3 人源性组织异种移植模型
Fig. 6-2-3 Patient-derived xenografts model

2.1.3 转基因小鼠技术的应用
2.1.3 Applications of Transgenic Mice Technique

利用转基因小鼠以及人源化小鼠技术构建的动物模型广泛应用于各种不同类型的肿瘤的研究，包括实体肿瘤（如肺癌、脑胶质瘤）和白血病。由于实验材料的限制，利用转基因小鼠构建人体肿瘤发生模型无疑可以极大地方便开展科学研究。PDX 模型不仅可以应用于基础医学研究，在临床上也有广泛的应用，如个性医疗、药效评价实验等。

Transgenic mice and PDX model are widely applied in tumor researching, such as solid tumor (lung cancer, glioma) and leukemia. The establishment of human tumorigenesis model by transgenic mice undoubtedly promotes the development of scientific research, considering the limitation of experimental materials. The PDX model can not only be applied in basic medicine, but also has wide application in clinic, such as personality medical treatment, drug effect evaluation experiment, etc.

2.2 肿瘤生物信息学
2.2 Tumor Bioinformatics

2.2.1 生物信息学概述
2.2.1 Background of Bioinformatics

如图6-2-4所示，生物信息学是一门交叉学科，它通过综合利用生物学、计算机科学和信息技术解读大量而复杂的生物数据，在大数据中寻找基因、蛋白和代谢物的关联，从多角度探究疾病的发生和发展。近年来，随着生物学研究的不断进展和重大突破，蕴含着分子序列、结构和功能的高通量的生物学数据越来越多，庞大的数据分析和解读也成为推动生物学发展的必要环节。肿瘤通常是由基因组的突变（点突变、插入突变、缺失突变），染色体易位和基因表达过程中的翻译后修饰造成的。肿瘤生物信息学将肿瘤的基因组、转录组、蛋白质组、表观遗传组和代谢组数据进行多组学多水平的整合，利用计算机的解读能力进一步探究肿瘤的发生机制。如今，全面整合的肿瘤生物信息学资源数据库都是免费开放的，这些高通量数据可以用于肿瘤的分类、诊断和临床结果预测。如癌症基因组图谱。

As shown in Fig. 6-2-4, bioinformatics is a cross-subject, which synthetically uses biology, computer science and information technology to interpret a large amount of complex biological data, seeks for the correlation of genes, proteins and metabolites, and explores the occurrence and development of diseases from multiple perspectives. In recent years, with the continuous progress and breakthroughs in biology, there are more and more high-throughput biological data containing molecular sequences, structures and functions, so that the analysis and interpretation of large and complex data has become a necessary part. Tumors are usually caused by genomic mutations (point mutation, insertion mutation, deletion mutation), chromosomal translocations and post-translational modifications during gene expression. The cancer bioinformatics integrates the data of tumor genomes, transcriptomes, proteome, epigenome and metabolome, and further explores the mechanism of tumor occurrence. Today, fully integrated databases of tumor bioinformatics are public and free to access, and these high-throughput data can be used for tumor classification, diagnosis and prediction of clinical outcome, such as the cancer genome atlas (TCGA).

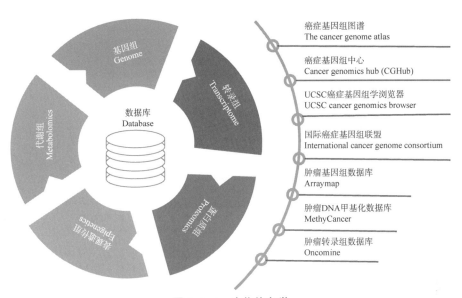

图 6-2-4 生物信息学
Fig. 6-2-4 Background of bioinformatics

2.2.2 肿瘤基因组图谱
2.2.2 The Cancer Genome Atlas

肿瘤基因组图谱是美国国家癌症研究所和美国人类基因组研究所共同合作的一个项目，旨在利用高通量的和多维度的数据更全面地解读肿瘤，以提高人们对肿瘤的认知和提高对肿瘤的预防、诊断和治疗能力。如图 6-2-5 所示，作为目前最大的肿瘤信息学数据库，TCGA 收录了胶质母细胞瘤、肺癌、卵巢癌和乳腺癌在内的 33 种癌症和 52 种癌型，每种癌型都包括基因表达数据、miRNA 表达数据、拷贝数变异、DNA 甲基化和单核苷酸变异数据。

The cancer genome atlas is a project that is mutually supervised by the U. S. National Cancer Institute and National Human Genome Research Institute. It aims at using high-throughput and multi-dimensional data to interpret tumors more comprehensively, and help people better understand, prevent, diagnose and treat cancers. As the largest tumor informatics database at present, TCGA shown in Fig. 6-2-5 includes 33 different kinds of cancers, such as glioblastoma, lung cancer, ovarian cancer, breast cancer and so on. It contains multiple data such as, gene expression, miRNA expression, copy number variation, DNA methylation and single-nucleotide polymorphism. The atlas of tumor shows the tumor mechanism of occurrence and development process.

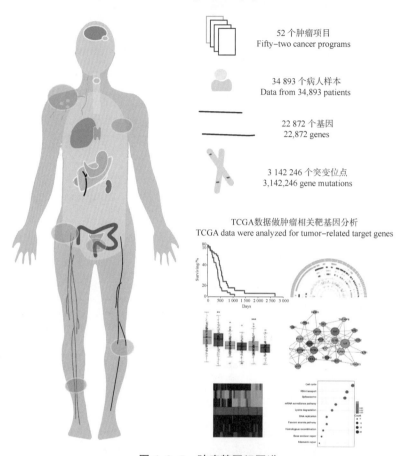

图 6-2-5 肿瘤基因组图谱

Fig. 6-2-5 The cancer genome atlas

2.2.3 泛癌图谱计划
2.2.3 Pan-cancer Initiative

泛癌计划（图 6-2-6）：2018 年《Cell》及其子刊《Cell Press》发布了泛癌症图谱相关的 27 篇高水平论文，泛癌图谱是 TCGA 的升级版，由美国国家卫生研究院资助的科学家团队完成的大规模癌症基因组图谱绘制。科学家对 TCGA 数据进行了充分的再解读和补充，从癌细胞起源、致癌过程和致癌途径三个角度诠释泛癌图谱，使人们重新审视肿瘤，更好地预防、诊断和治疗肿瘤。

In 2018, *Cell* and its sub-journal *Cell Press* published 27 high-level papers related to the pan-cancer atlas (Fig. 6-2-6), that is an updated project of TCGA and shows a large-scale cancer genome atlas and completed by a team of scientists funded by the U.S. National Institutes of Health. Scientists sufficiently reinterpreted and replenished to TCGA, and constructed the cancer atlas containing the origin, carcinogenesis and cancer pathway of cancer cells, which makes people re-understand tumor, and has the better prevention, diagnosis and treatment.

第6章 肿瘤治疗分析技术

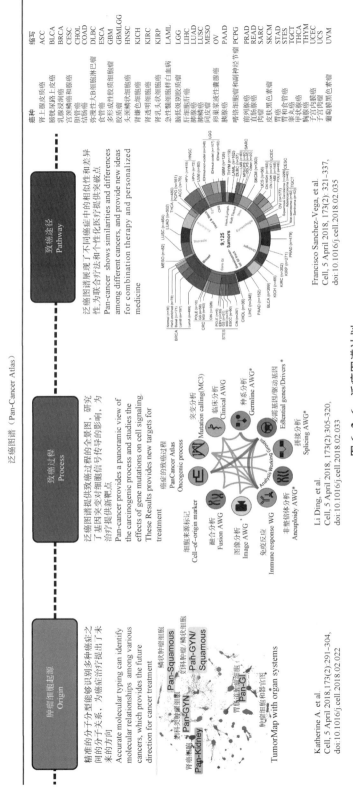

图 6-2-6 泛癌图谱计划
Fig. 6-2-6 Pan-cancer initiative

2.2.4 肿瘤单细胞测序技术
2.2.4 Single-cell Sequencing Technology

在肿瘤组织中,肿块中心的细胞、肿块周围的细胞,以及远端转移的细胞,其遗传信息存在较大的差异,这就是肿瘤遗传信息的异质性,这种现象在肿瘤中很常见。传统的测序方法是在多细胞水平进行,丢失了异质性的信息。单细胞测序技术从单细胞水平研究,使不同细胞类型得以精细区分,更精确地探究肿瘤的分子机制,如图 6-2-7 所示。

Heterogeneity between the cells is common phenomenon in tumor tissues, which makes the genetic information of the cells in the center of the tumor, around the tumor, and in the region of distal metastasis quite different from each other. The traditional sequencing method is carried out at the multicellular level, but the heterogeneous information is lost. Using single-cell sequencing technology, different cell types can be more exactly distinguished from the single cell level, and the molecular mechanism of tumor can be precisely explored, as shown in Fig. 6-2-7.

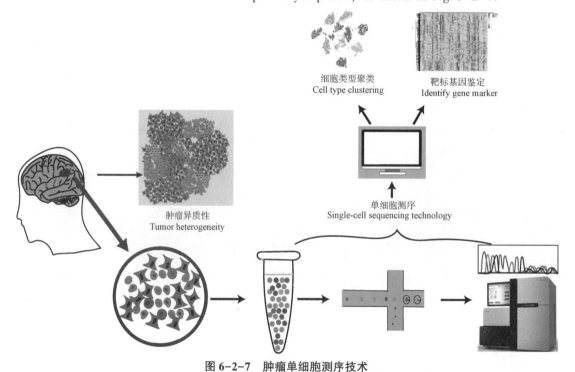

图 6-2-7　肿瘤单细胞测序技术
Fig. 6-2-7　Single-cell sequencing technology

Ⅲ 肿瘤的治疗
Ⅲ Treatment of Tumors

3.1 小分子靶向治疗
3.1 Small Molecule Targeted Therapy

3.1.1 间变性淋巴瘤激酶靶向药物
3.1.1 Targeted Drugs on Anaplastic Lymphoma Kinase

间变性淋巴瘤激酶（ALK）基因编码一种受体酪氨酸激酶（RTK），这种酶在大脑发育中起重要作用。目前，已经在间变性大细胞淋巴瘤、神经母细胞瘤和非小细胞肺癌中发现 ALK 突变、扩增或者重排，其中染色体重排最为常见，这种重排导致 ALK 与其他基因产生融合。比如在非小细胞肺癌中，2 号染色体倒位导致 EML4 和 ALK 两个基因片段融合并编码 EML4-ALK 融合蛋白，激活 RAS-MEK-ERK 和 PI3K-AKT 信号通路导致细胞增殖。作用于 ALK 基因的靶向药物（图 6-3-1）较多，如阿来替尼、哌柏西利等。但是，肿瘤细胞对 ALK 抑制剂也具有明显的耐药性，当其他信号通路上不同蛋白的基因发生突变时，肿瘤细胞的自我保护机制会促使旁路信号通路激活，从而不依赖原有的 ALK 信号通路进行增殖。肿瘤异质性也导致了肿瘤细胞的驱动基因可能不同，从而产生耐药性。因此研发联合用药治疗癌症尤为重要。

The anaplastic lymphoma kinase (ALK) gene encodes a receptor tyrosine kinase (RTK), which plays an important role in brain development. At present, mutation, amplification, or rearrangement of ALK have been found in anaplastic large cell lymphoma, neuroblastoma, and non-small cell lung cancer. Chromosome rearrangement is the most common phenomenon in cells, and leads to the fusion of ALK with other genes. For example, in non-small cell lung cancer, the inversion of chromosome 2 causes the fusion of EML4 and ALK gene fragments that encodes the EML4-ALK fusion protein. Consequently, this fusion protein activates the RAS-MEK-ERK and PI3K-AKT signaling pathways leading to cell proliferation. There are many targeted drugs (Fig. 6-3-1) on ALK, such as Aletinib and Pipercinil. However, tumor cells also have significant drug resistance to ALK inhibitors. When genes of different proteins in other signaling pathways are mutated, the self-protection mechanism of tumor cells will promote the activation of the bypass signaling pathway. As a result, tumor cells can proliferation without relying on the original ALK signaling pathway. Moreover, tumor heterogeneity also leads to the driving genes of tumor cells may be different and thus produce drug resistance. Therefore, it is particularly important to develop combined drugs to treat cancer.

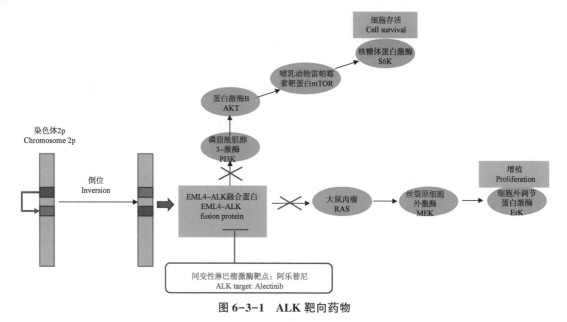

图 6-3-1 ALK 靶向药物

Fig. 6-3-1 Targeted drugs on ALK

3.1.2 mTOR 抑制剂
3.1.2 mTOR Inhibitor

PI3K/AKT/mTOR 是调节细胞周期的重要细胞内信号通路,与细胞的休眠、增殖、癌变和寿命直接相关。在多种癌症中,PI3K/AKT/mTOR 通路是过度活化的,由此减少细胞凋亡并促进增殖。mTOR 是一类丝/苏氨酸激酶,在细胞生长、分化、增殖、迁移和存活上扮演着重要角色。因此 mTOR 是抗肿瘤治疗的有效靶点。雷帕霉素(RAPA)是一种三烯大环内酯类的化合物,是经典的 mTOR 抑制剂(图 6-3-2),已被用于治疗多种恶性肿瘤,包括肾细胞癌、胰腺癌和乳腺癌。mTOR 容易受雷帕霉素抑制,通过调控细胞外信号,抑制 mTOR 信号,从而抑制肿瘤细胞增殖。RAPA 使 mTOR 信号通路下游靶蛋白磷酸化,参与调控蛋白质的合成;同时降低细胞周期蛋白依赖激酶(CDK)和细胞周期蛋白复合物激酶的活性来调控细胞相关进程,并且抑制细胞的增殖,达到诱导细胞凋亡的目的。RAPA 联合常规化疗药物一起作用于肿瘤时具有协同作用,临床应用价值高。

PI3K/AKT/mTOR is important intracellular signaling pathway that regulates the cell cycle and are directly related to cell dormancy, proliferation, cancer ation, and longevity. Among many cancers, the PI3K/AKT/mTOR pathway are highly activated, thus reducing apoptosis and promoting proliferation. mTOR is a type of serine/threonine kinases that plays an important role in cell growth, differentiation, proliferation, migration, and survival. Therefore, mTOR is an effective target for anti-tumor treatment. Rapamycin (RAPA) is a triene macrolide compound that is a classic mTOR inhibitor (Fig. 6-3-2) and has been used to treat a variety of malignant tumors, including renal cell carcinoma, pancreatic cancer, and breast cancer. mTOR is likely to be inhibited by rapamycin. By regulating extracellular signals, it inhibits mTOR signals and inhibits tumor cell proliferation. RAPA causes mTOR signal pathway downstream target protein phosphorylation and participates in regulating protein synthesis; additionally, the activity of cell cyclin-dependent kinase (CDK) and cell cyclin complex kinase was reduced to regulate cell-related processes and inhibit cell proliferation and apoptosis is induced. RAPA combined with conventional chemotherapy drugs have synergies when acting on tumors, and clinical application is of high value.

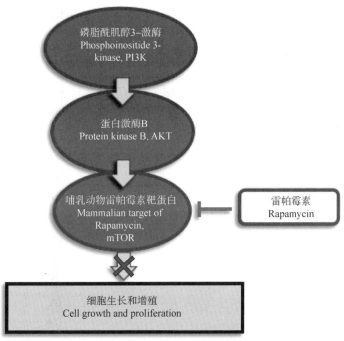

图 6-3-2　mTOR 抑制剂

Fig. 6-3-2　mTOR inhibitor

3.1.3　替莫唑胺
3.1.3　Temozolomide

恶性脑胶质瘤是成年人中最常见的中枢神经系统原发瘤。替莫唑胺（TMZ）（图 6-3-3）是恶性胶质瘤的化疗药物，具有较宽的抗肿瘤谱，与其他药物没有叠加毒性。TMZ 是一种 DNA 烷化剂，在体循环生理 pH 状态下，自发降解成一个开环结构的活性产物 MTIC［3-甲基-（三嗪-1-）咪唑-4-甲酰胺］。MTIC 会使 DNA 分子上鸟嘌呤和腺嘌呤甲基化。DNA 错配修复机制会把甲基化的碱基对应位置的核苷酸切除，并产生一个缺口，使 DNA 复制过程崩溃并使肿瘤细胞程序性死亡。随着时间的推移，脑胶质瘤细胞对 TMZ 引起的损伤产生了抵抗性。这个抗性的产生与多个机制有关，如 DNA 修复机制、表皮生长因子受体的过表达、p53 突变以及磷酸酶的表达等。进一步研究 TMZ 耐药性机制有助于找到导致肿瘤复发的新靶点，为肿瘤治疗提供新方案。

Malignant glioma is the most common central primary tumor of nervous system in adults. Temozolomide (TMZ) (Fig. 6-3-3) is a chemotherapeutic drug for malignant gliomas with a wide anti-tumor spectrum and no superimposed toxicity with other drugs. TMZ is a DNA alkylating agent that spontaneously degrades into a ring-opening structure of the active product MTIC (3-methyl-(triazine-1-) imidazol-4-formamide) at the physiological pH of the body cycle. MTIC causes guanine and adenine methylation on DNA molecules. The DNA mismatch repair mechanism removes nucleotides from the corresponding positions of methylated bases and creates a gap that collapses the DNA replication process and causes programmed tumor cell death. Glioma cells have developed resistance to TMZ-induced damage with time. The generation of this resistance is related to several mechanisms, such as DNA repair mechanisms, overexpression of epidermal growth factor receptors, and p53 mutations and phosphatase expression. Further research on TMZ resistance mechanism can help to find new targets for tumor recurrence and provide new solutions for tumor treatment.

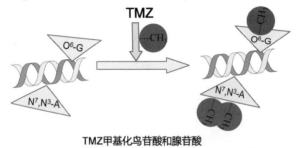

TMZ甲基化鸟苷酸和腺苷酸
TMZ methylated G and A

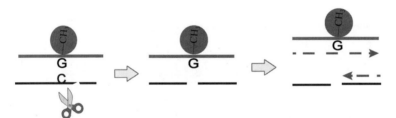

DNA自我修复机制切除甲基化的鸟苷酸对应链的碱基
The DNA repair mechanism removes the bases of the corresponding chains of methylated guanine

有缺口的NDA链复制时断裂
A chipped DNA chain breaks when it replicates

图 6-3-3　替莫唑胺

Fig. 6-3-3　Temozolomide

3.2　免疫治疗技术
3.2　Immunotherapy Techniques

3.2.1　免疫治疗技术分析
3.2.1　Immunotherapy Technique Analysis

目前的癌症免疫治疗技术主要分为三大类：过继细胞疗法、免疫检查点阻断剂和癌症疫苗，如表 6-3-1 所示。肿瘤免疫治疗是指通过激活机体的细胞免疫系统和体液免疫系统的内在能力，直接攻击肿瘤细胞，达到控制肿瘤发展和消灭肿瘤的目的。表 6-3-2 所示为三种免疫疗法的比较。

The current cancer immunotherapy technology is mainly divided into three categories, adoptive cellular immune therapy, immune checkpoint blockade, and cancer vaccines, as shown in Table 6-3-1. Tumor immunotherapy refers to the direct attack tumor cells by activating the intrinsic ability of the cellular immune system and the humoral immune system of the body to achieve the purpose of controlling tumor development and killing tumors. Table 6-3-2 shows comparison of three immunotherapies.

表 6-3-1 免疫疗法
Table 6-3-1 Immunotherapy

免疫疗法 Immunotherapy	原理 Principle
过继细胞疗法 Adoptive cellular immune therapy	通过向肿瘤患者输入在体外培养扩增或激活后具有抗肿瘤活性的免疫细胞，达到直接杀伤或激发机体免疫反应杀伤肿瘤细胞目的 The therapy of killing tumor cells directly or stimulating the body's immune response to kill tumor cells is to input immune cells with anti-tumor activity after in vitro culture amplification or activation to the people with cancer
免疫检查点阻断剂 Immune checkpoint blockade	免疫检查点是一系列调控免疫系统的抑制性信号通路。在肿瘤进程中，免疫检查点分子的活化和高表达能够抑制免疫细胞的功能，介导肿瘤免疫逃逸。利用相应的抗体如 CTLA4 抗体、PD1 抗体可以阻断和调控免疫检查点受体-配体的相互作用，逆转 T 细胞的免疫耗竭状态而发挥抗肿瘤效应 Immune checkpoints are a series of inhibitory signaling pathways that regulate the immune system. In the process of tumor, the activation and high expression of immune checkpoint molecules can inhibit the function of immune cells and mediate facilitate the immune escape of tumor. The use of corresponding antibodies, such as CTLA4 antibody and PD1 antibody, can block and regulate the interaction between receptor and ligand at the immune checkpoint, reverse the immune depletion of T cells and exert anti-tumor effect
癌症疫苗 Cancer vaccines	在体外大规模制备携带肿瘤相关抗原的细胞或被修饰过的肿瘤细胞作为治疗性疫苗对肿瘤患者进行治疗，激发患者机体针对肿瘤细胞的特异性免疫应答抑制，从而治疗肿瘤 Large-scale preparation of tumor-associated antigen-carrying cells or modified tumor cells in vitro as therapeutic vaccines to treat tumor patients and stimulate their specific suppression of the body's immune response against tumor cells to treat tumors

表 6-3-2 三种免疫疗法的比较
Table 6-3-2 Comparison of three immunotherapies

免疫疗法 Immunotherapy	优点 Advantages	缺点 Disadvantage
过继细胞疗法 Adoptive cellular immune therapy	HLA 非依赖的抗原识别、靶点类型多样、肿瘤识别特异性高 HLA-independent antigen recognition, diverse target types, high specificity of tumor recognition	对实体瘤的疗效仍不尽如人意；受到供体自身是否能够提供所需 T 细胞的限制；制备周期长且花费巨大；存在脱靶效应、细胞因子风暴、神经毒性和 B 细胞发育不全等治疗不良反应 The efficacy of solid tumors is still not satisfactory. It is limited by whether the donor can provide the required T cells. The preparation cycle is long and the cost is huge. There are adverse reactions such as off-target effect, cytokine storm, neurotoxicity and B cell dysplasia

续表

免疫疗法 Immunotherapy	优点 Advantages	缺点 Disadvantage
免疫检查点阻断剂 Immune checkpoint blockade	可以诱导晚期癌症患者产生更加有效和持续的应答，产生良好的治疗效果 Immune checkpoint blockade can induce a more effective and sustain response in patients with advanced cancer	有少数患者可以产生有效应答，部分患者在后续的治疗中产生耐药现象；免疫检查点阻断引起药物毒性称为"免疫相关的不良反应" Only a few patients can produce an effective response, and some patients develop drug resistance during subsequent treatment. In addition, the drug toxicity reaction caused by immune checkpoint blocking is called "immune related adverse events"
癌症疫苗 Cancer vaccines	不会损伤正常细胞；诱发免疫记忆细胞，维持长期的免疫效应，防止肿瘤和转移复发 Normal cells will not be damaged and can induce immune memory cells, maintain long-term immune effect, and prevent tumor and metastasis recurrence	肿瘤疫苗免疫原性较低，需要寻找合适的免疫佐剂联合使用以增加诱导机体免疫应答的成功率 Tumor vaccines have low immunogenicity, so it is necessary to find appropriate immune adjuvants to increase the success rate of inducing the immune response

3.2.2 嵌合抗原受体 T 细胞免疫疗法
3.2.2 Chimeric Antigen Receptor T-cell Immunotherapy

嵌合抗原受体 T 细胞免疫疗法（CAR-T）是一种治疗肿瘤的新型精准靶向疗法，如图 6-3-4 所示，近几年通过优化在临床肿瘤治疗上取得很好的效果。该疗法是肿瘤免疫疗法中的过继性细胞治疗的主要手段。CAR-T 治疗的主要过程是从癌症病人身上分离免疫 T 细胞，利用基因工程技术给 T 细胞加入一个能识别的肿瘤细胞，同时激活 T 细胞，杀死肿瘤细胞的嵌合抗体。此时的 T 细胞称为 CAR-T 细胞。随后体外培养扩增 CAR-T 细胞并输回病人体内。尽管 CAR-T 疗法已经取得了较好的临床效果，但是它还具有一定的风险。T 细胞在杀死其他细胞的时候会释放很多蛋白，产生炎症反应，使病人超高烧不退，甚至危及生命。

Chimeric antigen receptor T-cell (CAR-T) immunotherapy is a new precision targeted therapy for tumor treatment, as shown in Fig. 6-3-4. In recent years, it has achieved better results in clinical tumor treatment than before. CAR-T is the main method of adoptive cell therapy in tumor immunotherapy. The process of CAR-T treatment is to isolate immune T-cells from cancer patients. Using genetic engineering technology, T-cells are added with a chimeric antibody that can recognize tumor cells and activate T-cells to kill tumor cells. At this time, T-cells are called CAR-T cells at this time. CAR-T cells are then cultured and amplified in vitro and injected back into the patient. Although the treatment has achieved good results, it also carries certain risks, in which T-cells kill other cells and release proteins that trigger an inflammatory response that can leave patients with a high fever and life-threatening symptoms.

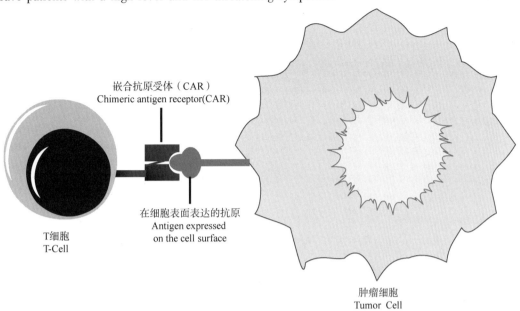

图 6-3-4 嵌合抗原受体 T 细胞免疫疗法

Fig. 6-3-4 Chimeric antigen receptor T-cell immunotherapy

第 7 章　生物信息技术

Chapter 7　Bioinformatics Technology

本章提要

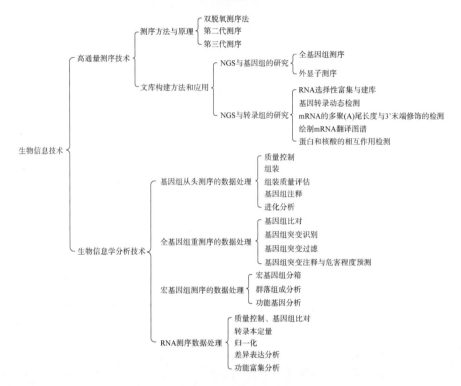

Summary of Chapter 7

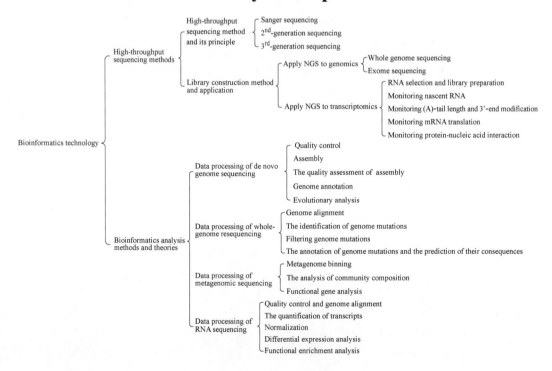

高通量测序技术
High-throughput Sequencing Method

DNA 测序技术作为分子生物学的核心技术，在生物医学的不同领域得到了应用，促进了其迅速发展。图 7-1-1 所示为测序的发展历程。双脱氧测序法（又名 Sanger 双脱氧链终止法）作为第一代非常经典且已经沿用了很长一段时间的测序方法，每次能够检测的 DNA 长度有限，一般只能用于单个基因的研究，无法满足全基因组水平的研究。为了适应这种需求，人们致力于发展新一代高通量测序技术。从 2004 年开始，陆续出现了以 Illumina Solexa、Roche Pyrosequencing、ABI SOLiD 为代表的第二代测序平台。这种技术进一步改良，可以做到同时检测几亿条 DNA 片段，从而大大降低了测序所需的时间和费用，解决了全基因组测序的难题。第二代测序平台中的序列信号是从每个单分子 DNA 扩增成一个 DNA 克隆后测定出来的，而 2008 年开始出现了以单分子测序为特征的 Pacific Biosorences SMRT、Oxford Nanopore、Helicos Helioscope 等第三代测序平台，这里的序列信号是直接从每个单分子 DNA 读出来的，不需要扩增成环节。相对于较为成熟的第二代测序平台，第三代测序技术还处在持续开发的阶段。新一代高通量测序技术的发展全面推动了不同层面宏基因组水平的研究，且在疾病预测和早期诊断上有广泛的应用前景。这个技术用于预测和诊断疾病的关键在于人们对特定基因和疾病之间的关系有了更明确的认识。

As the core technology of molecular biology, DNA sequencing technique has been applied to different fields of biomedicine, which promotes the rapid development of these fields. Fig. 7-1-1 shows the development of sequencing. As the 1st-generation sequencing method, dideoxy sequencing (also known as Sanger dideoxy chain termination method) is very classic and has been used for a long time. However, because it can only sequence the limited length of DNA at a time, it is only used for the study of a single gene, which cannot meet the growing need of genome-wide research. To meet demand, efforts are being made to develop the next-generation high-throughput sequencing technology. Since 2004, there have been several 2nd-generation sequencing platforms represented by Illumina Solexa, Roche Pyrosequencing and ABI SOLiD. This technology has been further improved to sequence hundreds of millions of DNA fragments simultaneously, which greatly reduce the time and cost of sequencing and solves the problem of whole-genome sequencing. The sequencing signals of the 2nd-generation sequencing platform arose from the DNA clones which are amplified from every single DNA molecule. And in 2008, the improved technology bring about a batch of new sequencing platforms characterized by single molecule detection. These 3rd-generation platforms such as Pacbio SMRT, Oxford Nanopore, Helicos Helioscope are able to read the sequence information directly from single DNA molecules without the step of amplification. However, compared with the mature 2nd-generation sequencing platform, the 3rd-generation technology is still in the stage of continuous development. The advent of the new generation high-throughput sequencing technology not only accelerates the research of metagenomics at different levels, but also has a wide application perspective in disease prediction and early diagnosis. The key to predict and diagnose diseases is to understand the relationship between specific genes and diseases, while the next generation sequencing technique can provide a better insight.

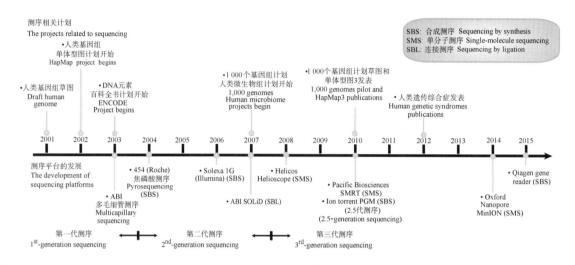

图 7-1-1　测序的发展历程

Fig. 7-1-1　The development of sequencing

高通量测序分为文库构建、利用高通量测序平台测序、生物信息学大数据分析三个部分。高通量测序平台的发展也促进了很多新方法学的出现。通过开发特殊的文库构建和数据分析的方法，它被广泛地应用到全基因组学、全转录组学和表观基因组学的研究。以下部分将围绕高通量测序平台的发展和原理、文库构建方法和应用、生物信息学分析方法和理论三个部分展开。

High-throughput sequencing is divided into three parts: library construction, sequencing using high-throughput sequencing platforms and bioinformatics big data analysis. The development of high-throughput sequencing platforms has also led to the emergence of many new methodologies. By developing special methods for library construction and data analysis, the new sequencing technology has been widely used in research of the whole genomics, transcriptomics and epigenomics. The following parts will focus on the development and principle of high-throughput sequencing platform, library construction method and application, and bioinformatics analysis method and theory.

1.1　高通量测序方法与原理
1.1　High-throughput Sequencing Method and Its Principle

弗雷德里克·桑格（Frederick Sanger）开发了名为双脱氧测序法（桑格测序法）的第一代测序技术，如图 7-1-2（b）所示。这个技术的主要原理类似于聚合酶链式反应（图 7-1-2（a））。聚合酶链式反应技术是 1983 年由美国化学家凯利·穆利斯开发的，原理十分简单，通过在体外模拟 DNA 聚合酶对模板 DNA 的半保留复制过程，对模板 DNA 进行指数级别的扩增。在这个反应中，需要加入待扩增的双链模板 DNA、分别和模板中靶序列两端互补的寡核苷酸引物、DNA 聚合酶和四种 dNTP。体外反应板体系中的双链模板 DNA 经过高温解链（变性），形成两条单链 DNA，降温退火后，引物和待扩增区域两侧的模板 DNA 链互补配对，以氢键形式结合到模板上。随后，体系中的 DNA 聚合酶开始沿引物 5′ 到 3′ 方向聚合 dNTP，延伸生成与模板互补的新生 DNA 链。如此反复 n 个循环，体系中的 DNA 分子会由 1 条变成 2^n 条。

Frederick Sanger developed the 1st generation of sequencing technology called dideoxy sequencing (Sanger sequencing) method, as shown in Fig. 7-1-2 (b). The main principle of this technique is similar to the polymerase chain reaction (PRC) (Fig. 7-1-2 (a)). Polymerase chain reaction was developed in 1983 by Kelly Mullis, an American chemist. The principle is very simple: by simulating the semi-reserved replication process of template DNA by DNA polymerase in vitro, the template DNA is amplified exponentially. In the reaction system, double-stranded template DNA, oligonucleotide primers, DNA polymerase, and four kinds of dNTPs are added. The double-stranded template DNA in vitro reaction system is unwound at high temperature (denaturation) to form two single-stranded DNAs. After cooling and annealing, the primers bind to the flanking template region through hydrogen bonds. The DNA polymerase in the system then begins to polymerize dNTP along the 5′ to 3′ direction of the primer, extending to form new DNA strands complementary to the template. And if you repeat these steps n times, you're going to get from one DNA molecule to 2^n DNA molecules in the end.

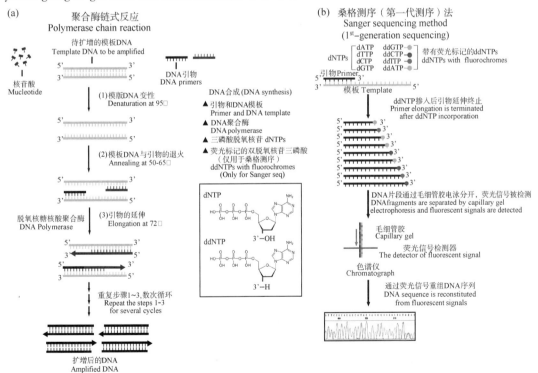

图 7-1-2 桑格测序法

Fig. 7-1-2 Sanger sequencing

(a) 主要原理；(b) 桑格测序法

(a) Main principle; (b) Sanger sequencing

桑格测序法是1977年提出的，比PCR技术要早。它经过不断改进，沿用至今。其工作原理与PCR技术类似，也是利用DNA聚合酶在体外复制DNA模板。而Sanger测序的巧妙之处在于反应体系中除了PCR反应中所需的酶、引物、模板、dNTP外，还按一定比例加入了四种带不同荧光标记的双脱氧核糖核苷酸（ddNTPs）。双脱氧核糖核苷酸3′端的羟基被氢取代，因此它一旦被DNA聚合酶聚合到新生链的3′端羟基上后，新生链就无法再延伸，复制终止。开始复制后，若模板DNA上第一个碱基是C，则其互补碱基是G，于是被加上的可能是dGTP或者带特定荧光团的ddGTP。如果是后者，则延伸结束，产生一个末端被荧光标记的短的DNA片段。若聚合的是dGTP，则延伸继续。在之后的每个位置，都面临着这样的两种可能性，其结果是会产生大量长短不一的带不同荧光标签的片段。当每个片段的大小只相差一个碱基时，可以按长度通过聚丙烯酰胺凝胶电泳分开。再根据每个长度所带的荧光种类，可以读出新生链序列。把这个序列反向互补一下，就得到了待测未知DNA模板序列。该技术经过发展，产生了Sanger自动测序仪。该仪器用毛细管凝胶电泳的方法，减少了所需样品量，激光激发后发出的荧光被检测器捕获，传入计算机，序列信息被自动读出。需要用凝胶电泳分离荧光标记DNA片段的Sanger测序法也被统称为第一代测序方法。其主要特点是读长可达1 000个碱基对以上，准确性高达99.999%，但测序成本相对较高，测序通量很低。

Sanger sequencing was proposed in 1977, earlier than PCR. It has been continuously improved and is still in use today. Its working principles are similar to PCR, which uses DNA polymerase to replicate DNA template in vitro. The magic of sanger sequencing is that in addition to the enzyme, primers, templates, and dNTPs as required for PCR, four dideoxyribonucleotides (ddNTPs) with different fluorescence labels are added to the reaction system in a certain proportion. The 3′terminal hydroxyl of the dideoxyribonucleotide is replaced by hydrogen, so once it is polymerized by DNA polymerase to the 3′terminal hydroxyl of the new strand, the new strand cannot be extended anymore and the replication stops. When replication begins, if the first base on the template DNA is C, then its complementary base is G, dGTP or ddGTP with specific fluorophore will be incorporated into the new strand randomly in the position. If the latter is the case, the extension ends, producing a short piece of DNA with a fluorescent tag. If dGTP is incorporated, the extension continues. At each subsequent position, there are two such possibilities, and the result is a large number of segments of different lengths with different fluorescent tags. When the size of each fragment varies by only one base, it can be separated by length using polyacrylamide gel electrophoresis. The sequence of the newly formed chains can be read out according to the fluorescence type of each length. By complementing the gel read out sequence in reverse, we get the DNA template sequence. After technical improvements, Sanger automatic sequencer was produced. It uses capillary gel electrophoresis to reduce the amount of sample required. While the fluorescence emitted after laser excitation is captured by a detector, the signals are transmitted to a computer, and the sequence information is automatically read out. Sanger sequencing, which requires the separation of fluorescent labeled DNA fragments by gel electrophoresis, is also known as the first-generation-sequencing method. Its main feature is that the reading length can reach more than 1,000 basepair and the accuracy is up to 99.999%, but the sequencing cost is relatively high and the sequencing throughput is very low.

在实施人类基因组计划的过程中，人们迫切感受到需要一种新的高通量测序的方法。于是，继 Sanger 测序法之后出现了第二代和第三代测序方法。这次革新的测序技术也被统称为高通量测序技术，如图 7-1-3 所示。新技术的主要标志是可以大规模平行处理多个样本，通量高，价格低，在正式进入功能基因组时代的今天，极大地满足了科研工作者的需求。它的基本思想是把无法直接测序的长基因组打碎成小的片段，通过测序这些短片段后，用生物信息学算法进行拼接组装，从而得到完整的基因组。

In the process of implementing the Human Genome Project, people felt an urgent need for a new high-throughput sequencing method. Thus, the 2nd and 3rd generation of sequencing methods appeared. And then, the innovative sequencing technology is called high-throughput sequencing technology, as shown in Fig. 7-1-3. The main feature of the new technology is that it can process multiple samples in parallel on a large scale, which enables its high throughput and low price. Today, when the era of functional genome is coming, the high-throughput sequencing technology greatly meets the needs of scientific researchers. The basic idea of the technique is to break up a long genome that cannot be sequenced directly into small pieces. And by sequencing these short pieces, bioinformatics algorithms can be used to assemble them to the complete genome.

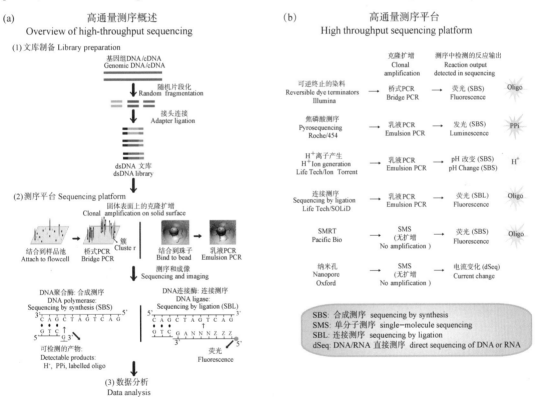

图 7-1-3　高通量测序技术

Fig. 7-1-3　High throughput sequencing

（a）高通量测序三个部分；（b）扩增

(a) Three parts of high-throughput sequencing; (b) Amplification

如前所述，高通量测序分为文库构建、利用高通量测序平台的测序、生物信息学大数据分析三个部分，如图 7-1-3（a）所示。文库构建的主要目的是把待测的基因组 DNA、逆转录得到的 cDNA 等样品进行碎片化（可通过超声波等机械打碎，也可通过转座酶、核酸内切酶等来完成），并在其两端连接特定序列的寡核苷酸接头。这段序列为后续的 PCR 扩增和测序提供引物的结合位点。根据需求接头也可以加一些不同的功能区域，如加条形码/索引序列，可以用于区分来自不同样品的文库。

High-throughput sequencing can be divided into three parts, library construction, sequencing using high-throughput sequencing platform and bioinformatics big data analysis, as shown in Fig. 7-1-3 (a). The main purpose of library construction is to fragment the genomic DNA, cDNA obtained by reverse transcription and other samples (which can be mechanically broken by ultrasonic waves, transposase, endonuclease, etc.). Then, the oligonucleotide (Adaptor or Linker) of specific sequences are linked at both ends of the DNA fragments. The adaptor sequence provides primer binding sites for subsequent PCR amplification and sequencing. To meet different needs, the added sequence can also include some different functional areas such as barcode/index sequence which can be used to distinguish the original library of different samples.

如图 7-1-3（b）所示，第二代高通量测序平台为了更准确地读取序列信号，首先把文库中的每个测序模板 DNA 分开并通过 PCR 进行扩增。扩增后每个模板 DNA 形成相同序列的 DNA 克隆，再从每个单克隆读取序列信号，以这种方式放大测序信号，达到降低读取错误率的目的。

As shown in Fig. 7-1-3 (b), in order to read the sequence signal more accurately, the 2nd generation high-throughput sequencing platform firstly separates the DNA sequencing templates in the library and amplifies them individually by PCR. After amplification, each template DNA forms a DNA clone of the same sequence, and then the sequence signal is read from each DNA monoclone. In this way, the sequencing signal is amplified, so as to reduce the reading error rate.

第二代测序平台分离"单模板"及"扩增"的方法可分为桥式 PCR（Bridge PCR，bPCR）和微乳液 PCR（emulsion PCR，ePCR）两种。桥式 PCR 在测序芯片（Flowcell）表面固定了与接头序列互补的引物。每个 DNA 片段经过 PCR 循环，最终每个单分子 DNA 会在其所在位置上扩增出很多拷贝并形成簇，从而达到测序时信号放大的功能。测序反应中加到测序芯片上的文库量需要严格控制。既要保证芯片上的簇之间相隔一定距离，不会在读取信号时相邻的簇之间互相干扰，同时也要避免形成的簇数目太少而导致最终得到的通量低。

The ways to amplify the library DNAs can be categorized into the Bridge PCR (bPCR) and the emulsion PCR (ePCR). Bridge PCR fixed primers complementary to the adaptor sequences on the Flowcell surface. Each DNA fragment goes through PCR cycles, and eventually each single molecule of DNA amplifies into many copies in its place forming a cluster, which amplifies the signal during sequencing. The amount of library added to the sequencing chip in the sequencing reaction needs to be strictly controlled. It is necessary to ensure that the clusters on the chip are separated by a certain distance, so that the adjacent clusters will not interfere with each other when reading signals. And at the same time, it is necessary to avoid forming too few clusters which will result in a low output.

另一种微乳液 PCR 的方法则在测序微珠表面固定了与接头序列互补的引物，通过油包水的乳化作用，形成每个油滴只包含一颗测序微珠和一个文库片段的结构，然后在油滴里面进行 PCR，单一的文库片段的 PCR 产物会被固定在测序微珠表面上，同样达到扩增信号的目的。这种方法也需要控制好测序微珠和上样的文库量，一个微珠和一个文库片段形成一个扩增油滴，降低测序错误率的同时，提高测序读取数量。

Another kind of signal amplification method, emulsion PCR, fixed primers on the surface of micro beads. Through a water-in-oil emulsion, each lipid droplet inside the oil contains only one sequencing microsphere and one DNA fragment from the library. PCR reaction is taken place in each of the microspheres and the products are fixed on the surface of the beads. This method also needs to control the number of sequencing microbeads and the amount of library. As one microsphere and one library fragment form an amplified oil droplet, the sequencing error rate has been reduced and the number of reads has been increased.

第三代测序平台不需要"单克隆"扩增阶段，直接采用单分子测序法（SMS）读取信号。

The 3rd-generation sequencing platform does not require the "monoclonal" amplification phase, directly uses single-molecular sequencing (SMS) to get the readable signals.

各平台根据测序反应的特点和读取的信号不同，分为不同的方法。根据反应的特点分为合成测序法（SBS）、连接测序法（SBL）和直接测序法（dSeq，见图 7-1-3（b））。这些方法都需要把化学反应的产物或待测核苷酸转换成机器可读的信号并检测。除了 Life Technology SOLiD 和 Oxford Nanopore，大部分测序平台采用合成测序法。它是利用聚合酶每次在测序链 3′端合成一个与模板链互补的核苷酸，并检测每次合成反应的产物。这些产物可能是带有荧光团的寡核苷酸、焦磷酸（PPi）或氢离子（H^+），它们会进一步被转化成机器可读的荧光信号、光信号和电信号。连接测序平台，如 ABI SOLiD 平台，则利用连接酶把测序链 5′端和带有荧光标记的探针寡核苷酸 3′端连接在一起。因为只有当探针的前两位与模板完全互补时，探针才能被连接到测序链上。与合成法类似，这里的连接产物也会被转化为荧光信号，转换成序列信息。直接测序法，如 Oxford Nanopore，检测当 DNA 或 RNA 单链通过嵌在人工膜的纳米孔蛋白时各个碱基形成的不同电流，可以根据不同的电流信号解读序列信号。

The platforms use different methods to detect the sequence information according to the chemical reactions and the signals read. According to the characteristics of the sequencing reaction, it can be classified into sequencing by synthesis (SBS), sequencing by ligation (SBL) and direct Sequencing (dSeq). These methods involve converting the products of chemical reactions or nucleotides to machine-readable signals and detecting them. With the exception of Life Technology SOLiD and Oxford Nanopore, most sequencing platforms employ synthetic sequencing. It uses polymerase to synthesize a nucleotide complementary to the template chain at the 3′ end of each sequencing chain, and detects the reaction product of each synthesis reaction. These products may be oligonucleotides with fluorescent labeling, pyrophosphate (PPi), or hydrogen ions (H^+), which are further converted into machine-readable fluorescence, luminescence and electrical signals. Ligation dependent sequencing platforms, such as ABI SOLiD platform, uses ligase to link the 5′

end of the sequencing chain to the 3′ end of the fluorescent labeled probe oligonucleotide. When the first and second positions of the probe are completely complementary to the template, the probe can be ligated to the sequencing chain. Similar to the synthesis method, the ligated products here are converted into fluorescent signals and sequence information. Direct sequencing methods, such as Oxford Nanopore, detect the different electric currents generated by each base when a single strand of DNA or RNA passes through a Nanopore composed of proteins embedded into the artificial membrane, and interpret sequence signals according to different current signals.

在二代测序平台中，利用最广泛的是 Illumina Solexa，如图 7-1-4 所示。Illumina 平台先用桥式 PCR 在测序芯片上形成数以亿计的单分子簇（图 7-1-4（a）），之后利用可逆终止的 DNA 合成化学反应原理对这些单分子簇同时进行测序（图 7-1-4（b））。这种方法使用带有不同荧光基团的四种特殊脱氧核糖核苷酸（dNTP）。荧光标记的 dNTP 3′位连接一个叠氮基团（阻滞基团），可以阻止下一个 dNTP 连接到测序链。但当有巯基试剂时，叠氮基团会发生断裂，并在 dNTP 3′位留下一个羟基，测序链重新获得延伸能力。所以在这个方法中 DNA 合成的终止是可逆的。同时，dNTP 中碱基与不同荧光分子也依靠叠氮链相连。每聚合一个荧光 dNTP-3′阻滞基团后，冲洗清除所有未反应的 dNTP，读其荧光信号，再用巯基化学试剂同时切除阻滞基团和荧光基团，暴露 3′羟基，再进行下一个位置碱基的聚合。这个方法的优点是快速、准确（错误率小于 1%，测序周期以人类基因组为例，30X 深度需要 1 周）；缺点是随着读长的增加，信噪比会下降，错误率会增加，所以每个簇测序长度比较短，只能达到几百个碱基。

Illumina Solexa is the most widely used 2nd-generation sequencing platform, as shown in Fig. 7-1-4. Illumina platforms use bridge PCR to form billions of single molecular clusters (Fig. 7-1-4 (a)) on sequencing chips. These single-molecule clusters are sequenced simultaneously using the principle of reversible DNA synthesis chemical reactions (Fig. 7-1-4 (b)). This method uses four special deoxyribonucleotides (dNTPs) labeled with different fluorescent groups. The fluorescent-labeled dNTP 3′ terminus is attached to an azide group 3′ blocker, preventing the next dNTP from joining to the sequencing chain. When a thiol reagent is present, the azido-bond breaks and leaves an hydroxyl group at dNTP 3′ terminus, allowing the sequencing chain to regain its ability to extend. Thus, the termination of DNA synthesis is reversible in this method. Meanwhile, the bases in dNTPs are linked to different fluorescent molecules by azido linkers. So after polymerizing a fluorescent 3′ blocker dNTP and then washing away all unreacted dNTPs, its fluorescence signal is read. Then, the blocker and the fluorescent group are removed with a sulfhydryl chemical reagent, exposing the 3′ hydroxyl group to polymerization at the next position. The advantage of this method is that it is fast and accurate (the error rate is less than 1%, and the sequencing takes 1 week at 30X depth for the human genome, for example). The disadvantage of this method is that as the read length increases, the signal-to-noise ratio decreases, leading to the increase in error rate. Therefore, the sequencing length of each cluster is relatively short, only reaching a few hundred bases.

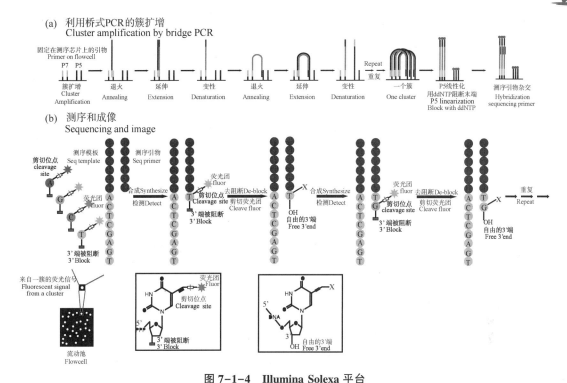

图 7-1-4 Illumina Solexa 平台

Fig. 7-1-4 Illumina Solexa Platform

（a）形成单分子簇；（b）测序

(a) Form single molecular clusters; (b) Clusters are sequenced

除此之外，Ion Torrent 及 Roche 454 测序仪分别捕捉聚合反应中释放的 H^+ 或焦磷酸，H^+ 引起 pH 值变化，焦磷酸转换成化学发光，如图 7-1-5 所示。Roche 454 测序仪的芯片微孔中固定 DNA 模板链，随后依次掺入 GCTA 四种脱氧核苷酸。如果有 DNA 聚合反应，释放出焦磷酸，通过化学反应生成 ATP，进一步进行化学发光反应并检测光信号。如果没有聚合反应，就接收不到光信号，继续循环加入另一种 dNTP。罗氏 454 测序仪因为没有市场竞争性，已经被淘汰。而另一种 Ion Torrent 的核心技术是使用半导体技术测定 H^+ 引起的 pH 的变化，它也在半导体芯片的微孔中固定 DNA 模板链，随后依次掺入 GCTA，聚合反应释放出的氢离子在穿过每个孔底部时能被检测到。这些方法使用天然核苷酸和聚合酶，测序成本相对低，但由于化学变化的强度不一定与聚合的碱基数目成正比，当含有连续的相同碱基的测序时，需要依赖算法校正。

In addition, Ion Torrent and Roche 454 sequesters capture pH changes or chemiluminescence respectively caused by H^+ or pyrophosphate released during polymerization reactions, as shown in Fig. 7-1-5. Roche's 454 sequencer fixes DNA templates on beads in the micropores and then adds GCTA four deoxynucleotides one by one. DNA polymerization reaction will produce pyrophosphates, then a chemical reaction turns them into ATPs which can go on to chemiluminescence generating light signals. If there is no polymerization, the light signal is not detected, and another type of dNTP is added in the cycle. Roche 454 sequencers have been phased out because there is no competitive market. Ion Torrent technology uses semiconductor technology to measure pH changes caused by H^+. It also immobilizes DNA templates on beads in the micropores of semiconductor chips, which are then mixed with one type of GCTA nucleotides in each round. The H⁺ ions released by the polymerization are detected as they pass through the bottom of each hole. These methods use natural nucleotides and polymerases so they are relatively inexpensive. However, as the strength of the sequencing signal is not necessarily proportional to the number of polymerized bases, algorithms are required to correct sequences containing successive identical bases.

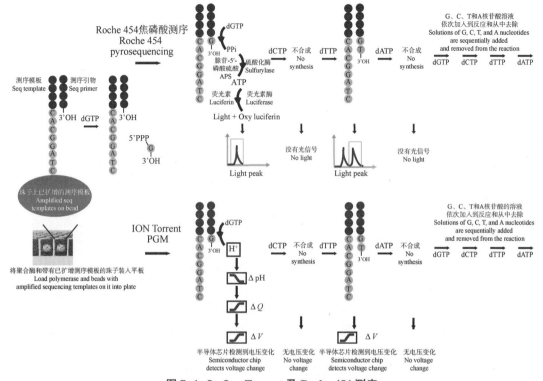

图 7-1-5　Ion Torrent 及 Roche 454 测序
Fig. 7-1-5　Ion Torrent and Roche 454 sequencing

第二代测序技术虽然成本低、快速准确、通量高，但一个很大的缺点是测序长度短，只能测到几百个碱基。这时，第三代测序技术，如 Oxford 的纳米孔单分子测序技术和美国太平洋生物技术公司（PacBio）单分子实时（SMRT）测序应运而生，如图 7-1-6 所示。Oxford 的纳米孔单分子测序在人工膜上嵌入纳米孔蛋白，并在纳米孔道形成电流（图 7-1-6（a））。当测序模板 DNA 双链分子结合机动蛋白，停靠到纳米孔时，DNA 双链分子解开，模板单链进入纳米孔。不同碱基通过纳米孔时对电流的阻遏不同，利用不同碱基引起的不同电流变化模式进行测序。PacBio 单分子实时测序，同样利用四色荧光标记 dNTP 的 DNA 合成来测序，但不经过 PCR 扩增放大荧光信号，而是采用零模波导孔（ZMW）来放大 dNTP 聚合的荧光信号（图 7-1-6（b））。SMRT 芯片上有 100 万个 ZMW 孔，单个 DNA 聚合酶固定在每个孔中。当一个单链测序模板通过 ZMW 孔时，荧光标记的 dNTP 聚合到测序模板链上，检测荧光信号，链继续延伸的同时荧光基团脱落扩散离开 ZMW 孔，从而使 ZMW 孔可以增强下一个聚合上的碱基的荧光信号。三代测序在测序长度上有不可替代的优势，提高了测序后基因组和转录组组装的准确度。但三代测序也有许多需要改进的地方，如准确率。因为是单分子测序，其信噪比会偏低。在通常情况下，改进的方法有优化测序酶、多次反复测同一片段及信号分析算法的优化。

Although the 2nd-generation sequencing is cheap, fast and accurate, with high throughput, it has a big disadvantage that the sequencing length is short, reaching only a few hundred bases. So, the 3rd generation of sequencing technologies, such as Oxford's nanopore single-molecule sequencing technology and PacBio's single-molecule real-time (SMRT) sequencing, have been developed, as shown in Fig. 7-1-6. Oxford's nanopore sequencer has nanopore proteins inserted into the artificial membrane and generates an electric current in the nanopore channels (Fig. 7-1-6 (a)). When the double-stranded template DNA binds to the motor protein and docks to the nanopore, the double-stranded DNA is released into two single-stranded DNA, and one strand enters the nanopore. Different bases make different current change when passing through the nanopore, and the different current patterns represent different DNA sequences. PacBio SMRT sequencing also employs the synthetic sequencing strategy of four-color fluorescence labeled dNTPs. Instead of PCR amplification, zero-mode waveguide (ZMW) is used to amplify the fluorescent signals during dNTP polymerization (Fig. 7-1-6 (b)). The SMRT chip has a million ZMW holes, each of which contains a single DNA polymerase. When a single chain of a sequencing template passes through the ZMW hole, the fluorescently labeled dNTP is polymerized to growing chain to detect the fluorescent signal. As the chain continues to extend, the fluorescent group falls off and is diffused away from the ZMW hole, so that the ZMW hole is ready for the next cycle of polymerization. The 3rd-generation sequencing has an irreplaceable advantage in sequencing length, which ensures the fidelity of genome and transcriptome assembly after sequencing. On the other hand, sequencing error rates still need to be improved. Because it is single-molecule sequencing, the signal-to-noise ratio is relatively low. In general, it can be achieved by optimizing the enzymes, repeatedly sequencing the same fragment several times and upgrading the base-calling algorithms.

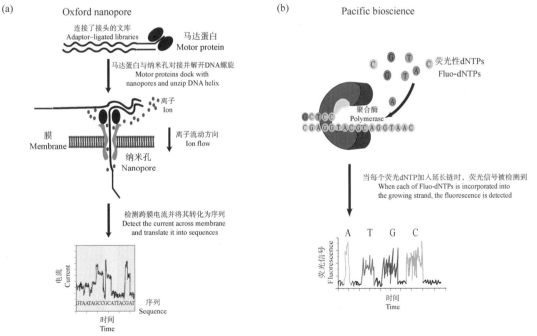

图 7-1-6 纳米孔单分子测序技术和单分子实时测序

Fig. 7-1-6 Nanopore single-molecule sequencing and single-molecule real-time sequencing

(a) Oxford 的纳米孔单分子测序；(b) PacBio 单分子实时测序

(a) Oxford's nanopore sequencer; (b) PacBio SMRT sequencing

高通量测序是一个划时代的技术。用第一代的最高性能测序仪测量人的基因组需要10年并耗费1.5亿美元,而用第二代代表性的 Illumina Hi-seq 技术只需1周、几千美元。测序技术的发展促进了全基因组学、全转录组学和表观基因组学的新研究方法的出现,也推动了遗传疾病的研究和早期医疗预测诊断。

High-throughput sequencing is an epoch-making technology. It takes 10 years and ＄150 million to measure the human genome with the highest-performance sequencer of the 1st generation, but now, only one week and a few thousand dollars with the 2nd-generation of Illumina Hi-seq. The development of sequencing technology has also promoted the emergence of new methods of whole-genome, whole-transcriptome and epigenome. Further, it has also promoted the early medical prediction and diagnosis of genetic diseases.

1.2 文库构建方法和应用
1.2 Library Construction Method and Application

下一代测序(NGS)技术的发展,促进了各物种全基因组的研究。为了达到在全基因组水平上检测基因表达的目的,人们一直在尝试开发新的方法。这些方法可以在全局上检测转录、mRNA 加工和降解、翻译以及蛋白质与核酸的相互作用,因此广泛应用于基因组学、表观基因组学和转录组学的研究。这些新方法的应用还开辟了遗传学、神经生物学和微生物学的新研究热点与领域。同时,随着新方法的出现,遗传病的致病基因鉴定也有了很大的进展,现在已经有约 6 000 个遗传病相关的约 4 000 个基因被鉴定出来,这些进展使疾病的预测和早期诊断成为可能。表 7-1-1 所示为分子机制已知的 OMIM 表型。表 7-1-2 所示为 OMIM 记录的疾病相关表型数及基因数。

The development of next-generation sequencing (NGS) technology has promoted the research on whole genome in various organisms. People have been developing novel of ways in order to achieve the purpose examining gene expression at a genome level. These strategies could globally monitor transcription, mRNA processing and degradation, translation, and protein-nucleic acid interaction, supporting the research on genomics, epi-genomics and transcriptomics. Applying these methods to genetics, neuronal biology and microbiology has opened new areas. The identification of genes which causes genetic diseases also has had a great advance after appearance of new methology. It has found about 4,000 genes are responsible for about 6,000 genetic disorders. The prediction and early diagnosis of diseases become feasible because of these advances. Table 7-1-1 shows OMIM phenotypes for which the molecular basis is known. Table 7-1-2 shows dissected OMIM morbid map scorecard.

表 7-1-1 分子机制已知的 OMIM 表型
Table 7-1-1 OMIM phenotypes for which the molecular basis is known

遗传方式 Inheritance pattern	2007.1 Jan. 2007	2013.7 Jul. 2013	2019.9 Sep. 2019
常染色体遗传病 Autosomal	1 851	3 525	5 252
X 染色体遗传病 X Linked	169	277	339
Y 染色体遗传病 Y Linked	2	4	5
线粒体遗传病 Mitochondrial	26	28	33
总计 Total	2 048	3 834	5 629

表 7-1-2 OMIM 记录的疾病相关系型数及基因数
Table 7-1-2 Dissected OMIM morbid map scorecard

表型分类 Class of phenotype	表型数 Phenotype	基因数 Gene
单基因病 Single gene disorders and traits	5 440	3 765
易患复杂疾病或感染 Susceptibility to complex disease or infection	693	501
非疾病 Nondiseases	147	116
体细胞遗传病 Somatic cell genetic disease	226	127

NGS 的基本应用是基因组的研究，如图 7-1-7 所示。为了从基因组中制备文库，基因组 DNA 首先被分割成短片段（TruSeq DNA 样本制备）。然后，gDNA 的单链区域被 DNA 聚合酶或核酸外切酶补齐或修剪成平末端。随后，双链的 gDNA 片段每条链的 5′OH 基团会被磷酸化，3′端羟基会被添加一个未配对的 dATP。然后通过碱基的互补配对，在 3′端有 T-突出碱基（未配对的 dTTP）的夹板接头序列会由 T4 连接酶连接到 gDNA 片段的两端。该文库随即可进行扩增和测序，以观察整个基因组或通过外显子富集的过程来检测基因组中由外显子组成的部分。利用基于微阵列芯片或溶液捕获的方法，可以实现特定基

因组区域的富集。在芯片富集法中，能够富集目标外显子区域的探针被固定在微阵列芯片的顶部，探针可以捕获外显子。在生物素探针富集法中，由于这些探针事先被生物素标记，它们连同被富集的外显子序列可被链霉亲和珠吸附并分离纯化。随后，富集库与探针分离并测序。外显子组包括大多数功能基因组。因此，Exome seq 是发现和研究同种生物的基因组差异的理想方法。

    The basic application of NGS is the research on genomes, as shown in Fig. 7-1-7. To prepare the libraries from a genome, genomic DNAs are first fragmented to short pieces (TruSeq DNA sample preparation). Then single-stranded end regions of gDNAs are blunted by fill-in and exonuclease activity. After phosphorylation of 5′OH group and addition of A-overhang at 5′-end, splint adaptors which have T-overhang at 5′-end are ligated to two ends of the gDNA fragments. This library can be amplified and sequenced to look at a whole genome (whole genome seq) or go through a process of exome-enrichment to examine the part of genome composed of exons (exome seq). The enrichment of specific genome region can be carried out by the method based on microarray-chip or in-solution capture. The probes which can enrich targeted exome regions are fixed on the top of a microarray-chip. Alternatively, the probes can be biotinylated and isolated by streptavidin beads after probes capture exome. The enriched library is separated from probes and sequenced. Exome is known to include most of the functional genome. So, Exome seq is an ideal way to find and study important differences of the genomes from the same species.

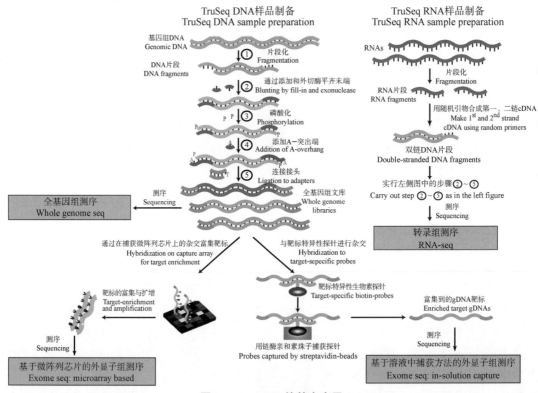

图 7-1-7　NGS 的基本应用

Fig. 7-1-7　The basic application of NGS

NGS 的其他主要应用是转录组的研究。通常，RNA 应该被反转录成 DNA，从而对它们进行测序。要从 RNA 中生成文库，有两种关键的反应：一种是接头序列的连接；另一种是使用引物在体外逆转录合成 cDNA。引物包括接头序列和随机序列。当起始 RNA 足够长时，可以使用 TruSeq RNA 样品制备。这一过程类似于从 gDNA 中制备文库，区别是前面增加了使用随机引物方法制造第一链和第二链 cDNAs 的额外步骤。

Other applications of NGS are the research on transcriptome. Usually RNAs should be converted to DNAs to sequence them. To make libraries from RNAs, there are two key reactions. One is adaptor ligation and another is synthesis of cDNA using primers which include adaptor sequences and random priming sequences. When the starting RNAs are long and enough, TruSeq RNA sample preparation can be used. The processe is similar to library preparation from gDNA, except for additional steps to make the 1st and 2nd strand cDNAs using random priming methods.

如果使用少量 RNA 生成 RNA-seq 库，可以选择 ScriptSeq 试剂盒。如图 7-1-8 所示，ScriptSeq 是一种基于逆转录酶和 DNA 聚合酶的标记系统，不同于传统的接头连接所依赖的末端标记方法。第一链 cDNAs 由逆转录（RT）合成，使用的引物包括随机六聚体 5′端标记接头（adaptor）序列。因此，每个随机引物合成的 cDNA 链都有一个 5′端标记接头。随后，为了给 cDNAs 添加 3′端标记，使用了随机五聚体序列的 3′端标记寡核苷酸（TTO）。这些寡核苷酸的 3′端都有阻遏基团。所以它们可以作为模板而不是引物。当寡核苷酸经随机五聚体区退火结合到 cDNAs 的 3′端时，DNA 聚合酶可将 3′端标记接头添加到 cDNAs 的 3′端。通过这种方式，可以获得双标记的 cDNAs 用于 NGS 测序。

If generating RNA-seq library using small amount of RNAs, ScriptSeq kit can be chosen. As shown in Fig. 7-1-8, different from conventional adaptor ligation-dependent terminal tagging method, ScriptSeq is a reverse transcriptase and DNA polymerase-based tagging system. The first strand cDNAs are synthesized by reverse transcription (RT) using primers which include random hexamer and 5′-end tagging (adaptor) sequences. So, the first strand cDNA synthesized by random priming has a 5′ tag (adaptor). In order to add a 3′-tag to cDNAs, terminal-tagging oligos (TTO) including random pentamer and 3′-end tagging sequences are used. These oligos have blocked 3′-end. So they can act as templates but not primers. When the oligos are annealed to the 3′-end of cDNAs by random pentamer region, 3′-tag could be added to 3′-end of the cDNAs by DNA polymerse. By doing that, di-tagged cDNAs are obtained for NGS sequencing.

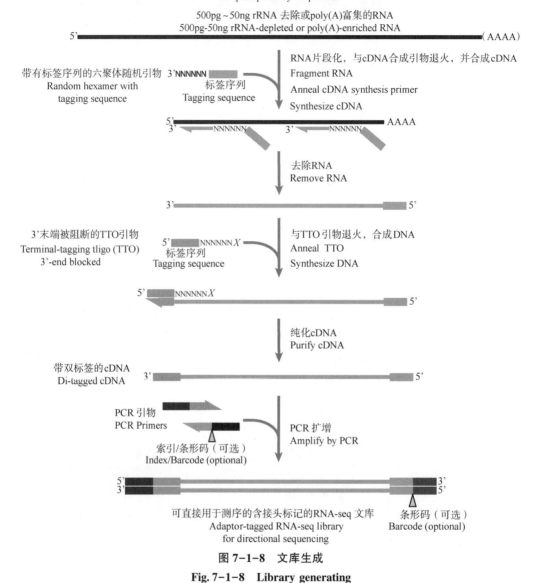

图 7-1-8 文库生成

Fig. 7-1-8 Library generating

为了检测特定类型的 RNA，在文库制备之前需要完成 RNA 的选择和纯化步骤。不同的 RNA 选择和去除方法都有各自的特点，因此可以得到多聚（A）尾的 RNA，不含 rRNA 的 RNA，或者只富集所需的转录本。表 7-1-3 所示为选择性富集 RNA 的方法。

To examine specific types of RNAs, the steps for RNA selection and purification should be performed ahead of library preparation. Different RNA selection and depletion methods have their own features so that people can get RNAs with poly (A) tails, rRNA-depleted, or only enriched in desired transcripts. Table 7-1-3 shows RNA enrichment methods.

表 7-1-3　选择性富集 RNA 的方法
Table 7-1-3　RNA enrichment methods

富集策略 Enrichment strategy	富集的 RNA 类型 Type of enriched RNA	核糖体 RNA 含量 Ribosomal RNA content	未剪接 RNA 含量 Unprocessed RNA content	基因组 DNA 含量 Genomic DNA content	富集方法 Enrichment method
总 RNA Total RNA	全部 All	高 High	高 High	高 High	无 None
多聚（A）尾 RNA 的富集 poly（A）selection	编码 RNA Coding RNA	低 Low	低 Low	低 Low	多聚（A）尾 RNA 与 oligo（dT）片段杂交 Hybridization with oligo（dT）probes
rRNA 的去除 rRNA depletion	编码，非编码 RNA Coding, noncoding RNA	低 Low	高 High	高 High	去除与探针互补的 rRNA Removal of rRNAs complementary to probes
特定 RNA 的富集 Desired RNA capture	目标 RNA Targeted RNA	低 Low	中 Moderate	低 Low	目标 RNA 与探针杂交 Hybridization with probes complementary to desired transcripts

传统的 RNA 测序技术只能检测 RNA 分子的丰度而不能研究其动态特性。因此，尽管这个结果能推断出蛋白质的丰度，但它忽略了基因调控的复杂性。这里，我们介绍三种常用来监测哪些是活跃转录下的 RNA 分子的方法，如图 7-1-9 所示。这些方法可以准确检测细胞在特定刺激（药物或环境因素）后的转录情况变化，对于研究基因表达的精细调控网络非常有用。

The conventional RNA seq technique detects the abundance of RNA molecules without studying the dynamic characteristics of them. Therefore, though the result infers the protein abundance, it overlooks the complexity of gene regulation. There are three frequently-used ways to determine what are the RNAs under active transcription, as shown in Fig. 7-1-9. These ways can accurately detect the cellular transcription shifts after certain stimulus (by drugs or environmental factors), which is ideal for understanding the fine regulation network of gene expression.

一种方法是基于细胞核 run-on 实验方法的技术。它让与 DNA 结合的聚合酶在被迅速分离出来的细胞核样本中的 DNA 链上滑动。因此，在溴化尿嘧啶（BrU）存在的情况下，体外的转录延伸反应会将 BrU 掺到新合成的 RNA 分子中。随后，总 RNA 被从细胞核中提取出来，再用抗溴化脱氧尿苷酸的抗体免疫沉淀出被 BrU 标记的 RNA 片段。测序这些纯化分离得到的 RNA 分子能提供 RNA 在 BrU 标记期间的合成和加工信息。这种体外检测方法的优点是可以保持较低的 RNA 降解，因此可以检测到一些不稳定的调控 RNA，如增强子 RNA。

The 1st method is based on nuclear run-on technique (run-on methods). It allows the DNA-bound polymerases to move through the DNAs in the quickly-isolated cell nuclei samples. Transcription elongation *in vitro* can happen in the presence of bromouridines (BrU), which incorporates BrUs into the newly-synthesized RNAs. Then, total RNAs are extracted from the nuclei, and the BrU-labeled RNAs can be purified by anti-bromodeoxyuridine antibodies (anti-BrdU) by immunoprecipitation. Sequencing these isolated RNAs can provide information regarding RNA synthesis and processing during BrU labeling period. The advantage of this in vitro method is that RNA degradation can be maintained low, so the detection of some unstable regulatory RNAs, for example enhancer RNAs, will be possible.

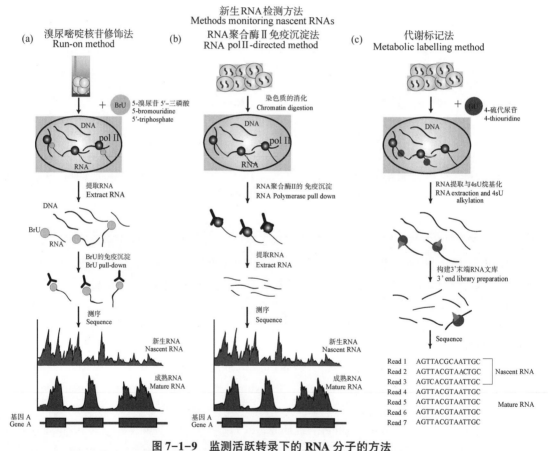

图 7-1-9 监测活跃转录下的 RNA 分子的方法

Fig. 7-1-9 Methods monitoring the actively transcribing RNAs

另一种 RNA 聚合酶 II 导向的方法。与 run-on 技术不同的是，提取细胞核后，核内染色体会被立即消化。由于在体内转录的 RNA 仍然与 RNA 聚合酶 II (RNAPII) 结合在一起，它们可以被抗 RNAPII 抗体拉下。接下来，当 RNA 提取步骤完成，这些分离得到的

RNA 片段将以标准的 RNA 测序方法进行建库和测序。该方法不像 run-on 方法那样依赖于体外延伸过程，它可以提供与转录同时发生的 RNA 加工情况。

The other method is RNA PolII-directed method. Different from the run-on technique, cell nuclei are extracted and the chromatins are digested immediately. Since the RNAs being transcribed in vivo are still associated with RNA polymerase II (RNAPII), they can be pulled down by anti-RNAPII antibodies. After RNA extraction steps, the selected RNA segments can then be sequenced with standard RNA seq procedure. This method doesn't rely on the in vitro process of elongation as the run-on method, and can provide information revealing the nature of co-transcriptional RNA processing events.

此外，还有一种新开发的方法叫作代谢标记（代谢标记方法）。该方法省略了细胞核提取和 RNAPII 复合物免疫沉淀的复杂实验过程，能提供转录合成和降解的信息。代谢标记的基本思路是在细胞培养液中加入核苷类似物 4-硫代尿苷（4sU）一段时间，细胞摄取 4sU，并在转录过程中 4sU 的加入可以标记新生的转录本。然后提取 RNA，通过碘乙酰胺（IAA）的亲核取代反应，在 4sU 标记的位置上共价修饰上羧氨甲基（carboxyamidomethyl-group）。在 4sU 烷基化后，从 RNA 提取物的 3′端逆转录，以胸腺嘧啶转化为胞嘧啶的方式构建 cDNA 文库。然后用 PCR 方法扩增 cDNA 文库，通过高通量测序方法以单核苷酸分辨率检测出 4sU 标记的新生 RNA 序列。

Besides, there is another new method called metabolic labelling (Metabolic labeling methods). This method omits the complex protocols of enucleation the cells and immunoprecipitation the RNAPII complex, providing information of transcript synthesis and decay. The basic idea of the metabolic labeling is to add 4-thiouridine (4sU), a nucleotide-analog, into the culture medium of cells for a certain time period, the 4sUs are immediately taken by the cells, and the incorporation of 4sU during transcription can therefore label the nascent transcripts. The RNAs are then extracted and the 4sU-labeled positions are covalently attached with a carboxyamidomethyl-group by nucleophilic substitution of iodoacetamide (IAA). After alkylation of the 4sUs, cDNA library is constructed by reverse transcription from the 3′ terminal of the RNA in a thymine-to-cytosine-conversions manner. The cDNA library is then amplified properly by PCR and the 4sU-labeled nascent RNAs are detected at a single-nucleotide resolution by the high-throughput sequencing.

PAL-seq 和 Tail-seq 可以全局地检测 mRNA 的多聚 A 尾长度及 3′末端的修饰，如图 7-1-10 所示。但目前的测序技术在长均聚物区域有很高的错误率。

PAL-seq and Tail-seq can examine poly (A) tail length and 3′ very-end modification on a global scale, as shown in Fig. 7-1-10. Current sequencing techniques have very high error rates in the region of a long homopolymer.

PAL-seq 与 Tail-seq 采用了不同的多聚（A）尾特征的检测策略。Tail-seq 方法使用不平衡的双端测序方法。在一对测序片段中，第一个测序片段可以明确转录本，而第二个测序片段则可以直接测出多聚（A）尾的序列。Tail-seq 依赖于一种解析同质性长序列的新算法——通过寻找荧光信号强度的拐点来解决长均聚物测序的难题，从而识别出 3′UTR 末端的多聚（A）序列的真正起始点。该方法的主要局限性是不能准确测量长度超过 231 个核苷酸或长度小于 8 个核苷酸的多聚（A）尾。这项技术的优势在于能够识别多聚尿苷酸和多聚鸟苷酸片段，这些片段有时会被加到转录本的 poly（A）序列后面，并可能在调控中发挥重要作用。与 Tail-seq 不同，PAL-seq 是基于单端读取的方法，从两个方面对序列进行分析。PAL-seq 不依赖于直接测序多聚（A）尾部，而是在测序芯片捕获文库并生成测序簇之后，测序引物结合到紧挨着多聚（A）尾部的转录序列的 3′端并启动延伸。第一步，在 PAL-seq 反应的前半部分，以多聚（A）tail 为模板，从引物序列的 5′端向 3′端延伸，此时只有 dTTP 和生物素标记的 dUTP 被加入反应体系中，随机掺入延伸产物。第二步，同一个方向接着进行标准 Illumina 测序，得到的 36 个核苷酸序列将用于进行转录本鉴定。在最后一步，生物素与荧光染料结合，荧光信号强度与生物素标记核苷酸的掺入量及多聚（A）尾巴的长度直接相关。理论上，PAL-seq 方法可以用来测量任意长度的多聚（A）尾，但这种方法也可能受到引物中生物素标记核苷酸的聚合效率的限制。

The differences between them relate to how the poly (A) tail is characterized. The Tail-seq protocol uses an unbalanced paired-end sequencing approach. The first read of the pair specifically identifies individual transcripts, while the second read of the pair is used to directly sequence the poly (A) tail. The Tail-seq approach relies on a novel solution to the challenge of sequencing through a long homopolymer by looking for an inflection in the fluorescent signal intensities to help identify the true start of the poly (A) sequence at the end of the 3′UTR. The primary limitation of this approach is that it cannot accurately measure poly (A) tails longer than 231 nucleotides or shorter than 8 nucleotides. This technique does offer the advantage of being able to identify uridine and guanosine stretches, which sometimes tail poly (A) sequences on transcripts and may have important roles in regulation. In contrast to Tail-seq, PAL-seq is based on a single-end read approach which is analyzed in two aspects. PAL-seq does not rely on directly sequencing the poly (A) tail. After libraries are captured on the flow cell and sequence clusters are generated, templates are primed at the 3′ end immediately adjacent to the poly (A) tail. The first step of the PAL-seq reaction uses the poly (A) tail region as a template to extend the primer sequence from the 5′ end meanwhile incorporating only dTTP and biotin-labeled dUTP into the reaction product. The second step uses the same primer for standard Illumina sequencing of 36 nucleotides next to the poly (A) tail region for transcript identification. In the final step, the incorporated biotin is conjugated with a fluorescent dye. Fluorescent signal intensity is directly related to tail length through the incorporation of the biotin-labeled nucleotides. In theory, the PAL-seq approach could measure poly (A) tails of any length, but be limited by the efficiency of biotin-labeled nucleotide incorporation.

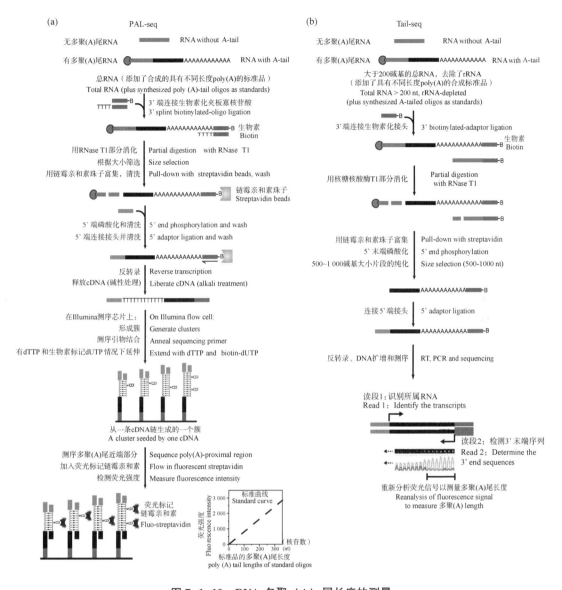

图 7-1-10　RNA 多聚（A）尾长度的测量

Fig. 7-1-10　Methods monitoring poly（A）tail length

核糖体在翻译阶段将 RNA 中的遗传信息解码为蛋白质。mRNA 翻译的调控在调节细胞内蛋白水平方面具有重要意义。精细调节的蛋白水平使细胞对快速变化的环境做出精确而迅速的反应。这里，我们介绍 4 种基于测序确定哪些 mRNA 分子正在被翻译的实验方法如图 7-1-11 所示。

Ribosomes decode genetic information in RNAs into proteins by translation. Regulation in mRNA translation is significant in tuning intracellular protein levels, which provides cells precise and prompt response to the rapidly changing environments. Here we described 4 sequencing-based approaches that can tell what mRNAs are translated, as shown in Fig. 7-1-11.

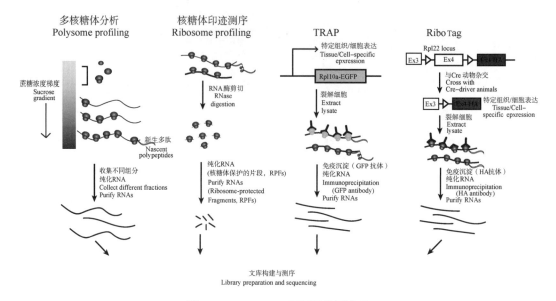

图 7-1-11　mRNA 翻译的检测方法

Fig. 7-1-11　Methods monitoring mRNA translation

最传统的方法是多核糖体分析。采用蔗糖梯度离心分离的方法分离出结合有不同数量的核糖体的 mRNA 组分。mRNA 上的核糖体越多，说明其翻译越活跃。分离得到的 mRNA 片段可以用来进行标准的文库构建和 RNA 测序。然而，蔗糖梯度离心很难分离出全长和足够数量的 mRNA。通过省略多聚核糖体分离步骤，核糖体印迹测序技术对多聚核糖体分析方法做了一些改进。它用低浓度的 RNase I 直接处理细胞裂解液，轻微消化其中的 mRNA 链。由于正在被翻译的 mRNA 序列结合了核糖体而被保护起来，这部分序列不会被消化，它们将被保留在裂解液中，然后提取被保护的 mRNA 片段并进行标准的文库构建和测序处理。这种方法比多聚核糖体分析更容易实现。但是，识别那些 mRNA 与更多核糖体结合的实验信息丢失了。因而，多核糖体分析会更依赖于生物信息学分析来获得准确的 mRNA 翻译图谱。

The most conventional method is polysome profiling. It employs the method of sucrose gradient centrifugation to isolate mRNA fractions bounding with different number of ribosomes. The more ribosomes are on a mRNA indicates the more actively translated. The isolated mRNA fractions can be conducted standard library construction and RNA-seq. However, it is difficult to isolate mRNAs of full length as well as enough amounts after the sucrose gradient centrifugation. Ribosome profiling is improved from polysome profiling by omitting the step of polysome isolation. It simply treats the cell lysate with low concentration of RNase I to digest the mRNAs partially. Since the ribosome-bound mRNA fragments that are under translation are protected from being digested, they can be then extracted for the standard library construction and sequencing process. This method is much easier to implement than polysome profiling, however, the experimental clues for discerning what mRNAs are bound by more ribosomes are lost. Therefore, polysome profiling depends on bioinformatic analysis to get the accurate map of mRNA translation.

TRAP 和 RiboTag 都利用免疫沉淀纯化被标记的核糖体-mRNA 复合物。这两种方法的主要区别在于标记核糖体的实验策略上。在 TRAP 技术中，核糖体蛋白 L10a 与 EGFP 编码区融合，在适当的强启动子的驱动下，表达形成 EGFP 标记的核糖体。而在核糖体标记实验方法中，则需要构建带核糖体标记的转基因小鼠模型。核糖体标记小鼠品系带有一种修饰过的核糖体蛋白 L22（Rpl22）。这种小鼠的 Rpl22 基因位点的最后一个可编码外显子 4 的两侧插入了一对同向的 LoxP 重组位点，其后再插入一个相同的外显子 4，这个外显子 4 的后面，在最后的终止密码子之前再插入血凝素（HA）标签。将构建好的核糖体标记小鼠与 Cre 驱动小鼠进行杂交，由于 Cre 重组酶是由组织特异性启动子驱动的，因此它在感兴趣的细胞类型中才能表达，从而切除两个 LoxP 位点之间的野生型 Rpl22 外显子 4。在切除野生型 Rpl22 外显子 4 后，Rpl22-HA 蛋白被翻译并形成带 HA 标签的核糖体。核糖体被标记后，利用各自的抗体（TRAP 为抗 GFP 抗体，核糖体标记为抗 HA 抗体）通过亲和免疫沉淀纯化，然后从分离的核糖体中分离提取 mRNA 并进行标准的 RNA 测序流程。这两种方法的优点是可以从复杂组织中获得特定细胞类型的 mRNA 翻译信息，缺点是核糖体蛋白的异源过表达可能改变细胞的正常生理状态，从而扭曲 mRNA 的翻译。此外，动物模型的建立会花费大量的时间和金钱成本，从而限制这些方法的广泛使用。

TRAP and RiboTag are two similar methods that use immunoprecipitation to purify the tagged ribosome-mRNA complex. The difference of these two methods is the way to tag the ribosomes. In TRAP technique, ribosome protein L10a is fused with EGFP coding region under the drive of a proper strong promoter which results in EGFP-labeled ribosomes. While in RiboTag method, ribosome-labeled transgenic mouse is constructed. The RiboTag mouse method has a modified Ribosomal Protein L22 (Rpl22). The final coding exon4 of Rpl22 gene loci is flanked by LoxP recombination sites followed by an additional exon4 that has hemagglutinin (HA) epitope coding sequences inserted before the stop codon. The RiboTag mouse is then crossed to a Cre driver mouse. Since the Cre recombinase is driven by tissue specific promoter, it is expressed only in the cell type of interest, resulting the removal of LoxP-flanked wild type Rpl22 exon4. After the excision of the wild type Rpl22 exon4, the Rpl22-HA protein is expressed, forming the HA-tagged ribosomes. Following the labelling of ribosomes, they are purified by affinity immunoprecipitation using the respective antibodies (anti-GFP for TRAP, anti-HA for RiboTag). Then, the mRNAs are extracted from the immno precipitate and standard RNA seq protocol is carried out. The advantage of these methods is that they can obtain the mRNA translation information of specific cell types. However, the heterogenous expression of ribosome proteins may change the normal physiological state of the cells, and thus distort mRNA translation. Also, the genetic manipulation of animals is time and money consuming which may limit the wide use of these methods.

蛋白质与核酸的相互作用可分为两种类型：DNA-蛋白质相互作用和 RNA-蛋白质相互作用，如图 7-1-12 所示。它们在调节不同水平的基因表达过程中发挥着关键作用。转录水平的基因表达调控主要依赖于特定的 DNA 序列与 DNA 结合蛋白之间的相互作用。目前最流行的一种方法是利用 Chip-seq 技术在整个基因组中精确定位感兴趣的蛋白质的结合位点。基因组 DNA（gDNA）及其结合蛋白首先通过低浓度的多聚甲醛化学交联形成共价键。然后，基因组被打碎，由此产生的 DNA 片段——蛋白质复合物将被 DNA 结合蛋白的特异性抗体纯化。DNA 通过逆向交联从纯化出来的复合物中分离出来，并通过高通量测序确定与蛋白结合的 DNA 区域。

There are two types of protein-nucleic acid interactions, DNA-protein interaction and RNA-protein interaction, as shown in Fig. 7-1-12. These interactions play key roles in regulating gene expression at different levels. Transcriptional regulation majorly depends on the interactions of specific DNA regions and DNA-binding proteins. The most popular method today to map the genome wide DNA-protein interactions is Chip-seq technique. The genome DNA (gDNA) and proteins are first crosslinked by low concentration of paraformaldehyde. Then, gDNA is fragmented by ultrasonication. The resulting DNA fragment-protein complex is immunoprecipitated by antibodies against the protein of interest. DNAs are extracted from the complex after reverse crosslinking, and sequenced to determine the exact protein binding regions.

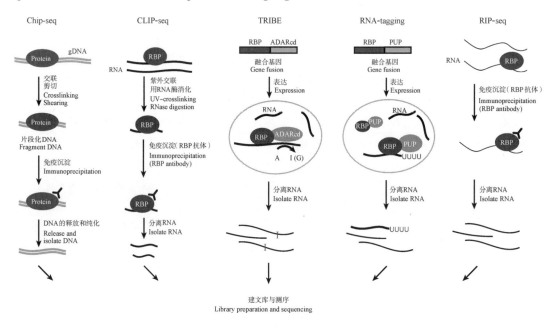

图 7-1-12　蛋白和核酸相互作用检测方法

Fig. 7-1-12　Methods monitoring protein-nucleic acid interaction

转录后基因表达调控很多是由 RNA-蛋白质相互作用所介导的。检测细胞内 RNA 与 RNA 结合蛋白（RBP）的结合位点的方法可分为两类：免疫共沉淀型和直接在细胞内标记靶标型。其中 RIP-seq 和 CLIP-seq 需要利用 RBP 的抗体从细胞裂解液中纯化与 RBP 结合的 RNA，然后对其进行高通量 RNA 测序。这两种方法的区别在于 CLIP-seq 在亲和纯化 RBP-RNA 复合物之前还有额外的步骤。CLIP-seq 的第一步是将活细胞暴露在一定剂量的紫外线下，使 RBP 与其靶 RNA 上的结合位点相互交联。CLIP-seq 的另一个额外步骤是用

RNase 部分消化细胞裂解液中的 RNA，只保留被 RBP 保护的 RNA 片段。CLIP 中的交联步骤可以促进捕捉不稳定、瞬时的 RNA-RBP 相互作用，提高检测的灵敏度，而 RNase 消化能提高相互作用位点的检测分辨率。虽然 CLIP-seq 能最直接地揭示体内 RNA 和 RBP 互作情况，且仍是目前确定 RBP-RNA 相互作用的金标准，但是它的抗体依赖亲和纯化过程以及低效的紫外交联过程（通常是 1%～5%）使其所需的起始样品量多，背景噪音高，不适用于检测少量特定细胞中的 RNA-蛋白相互作用。

Post-transcriptional regulation of gene expression islargely mediated by RNA-protein interactions. Methods used for detecting the RNA-protein interactions can be categorized into two types, the coimmumoprecipitation based method and the in-cell-target-modification based method. RIP-seq and CLIP-seq both use antibody against RBP to isolate the RBP bound RNAs from cell lysate, and then the RNAs are extracted from the complex and sequenced by next-generation sequencing technology. The differences between these two methods are that CLIP-seq has additional steps before the affinity purification of RBP-RNA complex. The first step of CLIP-seq is to expose living cells to a certain dosage of UV light, allowing the crosslinking between RBP and its target RNAs. Another extra step of CLIP-seq is the partial digestion of the cell extract with RNase, leaving only the RNA fragments which are protected by the RBP. The crosslinking step in CLIP can stabilize the weak and transient RBP-RNA interactions, increasing the sensitivity of detection. Further, the RNase digestion ensures the high resolution in mapping the exact RNA region bound by RBP. Although CLIP-seq reveals the in vivo interactions of RNA and RBP in a straightforward way, and is served as a gold standard for determination of RBP-RNA interactions, the requirement of affinity purification and inefficient UV-crosslinking (usually 1%～5%) make it need a large amount of starting materials and high background noise.

另外两种方法以体内标记靶标的方式检测 RNA-RBP 相互作用。其中，TRIBE 在细胞中表达感兴趣的 RBP 与 RNA 编辑酶 ADAR 催化结构域（ADARcd）的融合蛋白。RBP 引导融合蛋白结合至靶 RNA 位点，ADARcd 的部分会编辑结合位点相邻的 RNA，将其中的腺苷转化为肌苷。产生的编辑位点将通过标准的高通量 RNA 测序流程检测出来，这为 RBP 目标位点测定提供了依据。重要的是，ADARcd 的编辑是不可逆转的，因此能保证 TRIBE 方法检测的稳定性和准确性。此外，通过在融合蛋白中引入细胞特异性启动子，TRIBE 可适用于检测特定细胞中的 RBP-RNA 相互作用。类似于 TRIBE 的概念，RNA-tagging 方法融合表达感兴趣的 RBP 和源于线虫的 PolyU 聚合酶（PUP）。由于 PUP 的功能是将多个尿苷添加到 RBP 结合的 RNA 的 3′端，因此可以通过高通量双端测序检测 RBP 靶的 RNA。RNA-tagging 技术在构建 cDNA 文库时，通过"尿苷选择性"引物特异性地扩增 3′端 polyU 标记的 RNA，提高了方法的灵敏度。此外，RNA-tagging 技术用成对的双端测序方法，将 polyU 标记的 RNA 比对到基因组上。

The other two methods detect RBP-RNA interactions in an in-vivo-labeling manner. Targets of RNA-binding proteins identified by editing (TRIBE) expresses the fusion protein between RBP of interest and the catalytic domain of the RNA editing enzyme ADAR (ADARcd). The RBP part will guide fusion proteins to the target RNA regions, and after binding, the ADAR part will edit the RNAs near the binding sites, converting adenosines into inosines. The editing events will be determined by the standard RNA-seq procedure, providing information on the target recognition of RBP. Importantly, the editing by ADARcd is irreversible, ensuring the stability and accuracy of TRIBE. In addition, by introducing cell type specific promoters to the fusion protein, TRIBE can be used to detect RBP-RNA interactions in specific cell types. Similar to the concept of TRIBE, the RNA-tagging method makes a fusion protein between a RBP and the PolyU polymerase (PUP) enzyme from C. elegans. Since the function of PUP is to add uridines to 3′ ends of the RBP bound RNAs, the target RNAs can be detected by high throughput paired-end sequencing. RNA-tagging selectively amplifies the polyU labeled RNAs by "U-selective" primers when constructing cDNA libraries, which increases the sensitivity of the method. Further, RNA-tagging technology needs paired-end sequencing to map the polyU labeled RNAs back to their genomic positions.

基于高通量测序的新方法学也推动了相应生物信息学数据处理和分析方法的开发。

Novel methodology in NGS leads the development of bioinformatics tools for data processing and analysis.

Ⅱ 生物信息学分析方法和理论
Ⅱ Bioinformatics Analysis Methods and Theories

2.1 基因组从头测序的数据处理
2.1 Data Processing of De Novo Genome Sequencing

基因组从头测序是指对新物种或缺少参考基因组的物种进行高通量测序，并使用相关软件对未知基因组进行组装和功能注释，从而获得该物种的全基因组序列、基因组成以及进化特征等信息。目前该技术被广泛应用于建立不同物种的参考基因组。基因组从头测序的分析主要包含质量控制、组装、组装质量评估、基因组注释和进化分析过程。

Genome de novo sequencing refers to the process of high-throughput genome sequencing, genome assembly and functional gene annotation of new species or such species that lack reference genome by utilizing corresponding software. Such software is widely used to establish the reference genome of different species, provide information about the whole genome sequences, gene composition and evolutionary characteristics of species. The data process of genome de novo sequencing mainly includes the following processes: quality control, assembly, quality assessment, genome annotation and evolutionary analysis.

A. 质量控制

A. Quality control

测序完成后，从服务器下载测序结果。测序结果的原始数据是以 FASTQ 格式存储的文件，包含了测序仪检测到的所有 DNA 序列和测序质量信息。FASTQ 格式文件以一个测序读段为基本单位，记录从一条 DNA 片段读取的测序结果。每个测序读段由四行字符串组成，如图 7-2-1 所示。第一行字符串是以"@"符号开始，用于描述该序列的一些测序信息，如测序设备，测序读段 ID，以及是否为双末端测序等。第二行字符串是通过测序得到的 DNA 序列，由 A、T、C、G、N 五种符号组成，N 代表无法检测的碱基。第三行字符串通常只有一个"+"号，也可在其后面添加读段相关的描述信息。第四行字符串用于描述第二行测序序列的测序质量结果，这一行中的碱基测序质量值与第二行中的碱基一一对应。此外，碱基测序质量值（Q）与碱基测序错误率 P 之间的关系可用公式 $Q=-10\lg P$ 来表示。例如某个碱基的质量值是 20（$Q20$），那么通过上述公式计算得到该碱基被错误识别的概率为 1%，即正确率为 99%。在对测序结果进行分析之前，需要对测序得到的序列进行质量检测和过滤。常见的质量控制软件有 FastQC、Trimmomatic、Fastp 等。通过质量控制过滤，可去除测序结果中的接头序列、含有较多不确定碱基"N"以及平均测序质量 Q 值较低的测序读段。

After sequencing is complete, the results are downloaded from the server. The raw data of sequencing are stored in FASTQ format, which contains all DNA sequences and the quality information of sequencing detected by the sequencer. The basic unit of FASTQ format file is the sequencing read, which records sequencing information from each DNA fragment. Each sequencing read consists of four rows of data, as shown in Fig. 7-2-1. The first row starts with the symbol "@", which usually indicates sequencing information, such as sequencing equipment, ID of the read, and whether it is paired-end sequencing. The second row is the sequence obtained previously, composed of "A, T, C, G, N". N represents an undetermined base. The third row either only consists of a "+" symbol, or adds a related description of the read behind the "+". The fourth row describes the sequencing quality results of the second row. The bases in the second row are in exact correspondence with the base sequencing quality values in the fourth row. The relationship between the base sequencing quality value (phred quality score, denoted by Q) and the base sequencing error rate P is shown by the formula $Q=-10\lg P$. For example, if the quality value of a base is 20 ($Q20$), then according to the formula given, the error rate P of the base is 1% and the accuracy rate is 99%. Before analyzing the sequencing results, the sequences obtained need to undergo quality assessments and filtering. Quality control software includes FastQC, Trimmomatic, Fastp and others. Through quality control processing, the adapter sequence in the sequences will be removed, reads containing lots of uncertain bases "N" or with low average sequencing quality are removed as well.

```
@SRR3177714.1 HWI-ST1167:91:C4BMKACXX:1:1101:1232:1981 length=51
TNTTATTTCTCATAATATAGGCCGACGAGACATAGATTTTGTACCAGTCTT
+SRR3177714.1 HWI-ST1167:91:C4BMKACXX:1:1101:1232:1981 length=51
<#1=DDDDHHDHH@ GBGGGB<? FHHIB8? D@?? BFFAFFHI9D*? FF4CED
```

图 7-2-1 FASTQ 格式文件

Fig. 7-2-1 FASTQ format file

B. 组装

B. Assembly

组装的目的是从原始测序短序列中组装出更长的叠连群。目前基因组组装的策略主要有两种：第一种称为OLC，该方法基于测序序列之间的重叠关系，适用于组装早期通过Sanger测序所得到的少量、读段长的序列，代表软件有Phrap、CAP3等。随着二代测序技术的发展，测序数据量不断增加，读段变得更短，组装所需的计算机资源要求更高，基于读段重叠的方法已不能满足组装的需要。因此出现了第二种组装策略，即基于k-mer的de Bruijn图方法。De Bruijn图方法首先将读段切成长度为 k 的片段，基于这些k-mer的组合，构建de Bruijn图，然后在已构建的碎片化序列基础上，选择合适的片段，构建完整的基因组序列，代表软件有SOAPdenovo、velvet、ABySS等。基因组组装完成后，会得到一个FASTA格式的文件，里面包含不同长度的叠连群。每一条叠连群包含两部分信息，分别是该叠连群的标识符（UID），以及组成该叠连群的碱基序列。

The purpose of assembly is to construct long contigs from short sequences, using one of the two main strategies currently available for genome assembly. The first strategy, overlap-layout-consensus (OLC), is based on the overlap between the short sequences. This method has been used for assembly of results with the limited sequencing output and long read length, such as those from Sanger sequencing. Phrap and CAP3 are representative software applying OLC, used for early bioinformatics analysis. With the development of the 2nd-generation sequencing technology, the size of sequencing data continuously increases. Sequencing reads are also shorter, thus, computer resource requirements for assembly are higher than those that were sufficient in OLC. This means that assembly requirements cannot be met through OLC. Consequently, the 2nd-generation assembly strategy, relying on k-mer's de Bruijin graph, is introduced. The de Bruijn graph method first cuts the reads into segments of k-nucleotides (called k-mers), and then builds the de Bruijn graph based on the combination of these k-mers. From fragments that have been constructed, the program selects the appropriate fragments to generate the entire genome sequence. Currently, software employing the de Bruijn graph method includes SOAPdenovo, velvet, ABySS, etc. The above process produces a file in FASTA format, which lists different lengths of assembled contigs. Each contig consists of two parts, the unique identifier (UID) and the base sequence of the contig.

C. 组装质量评估

C. The quality assessment of assembly

完成基因组的组装后，需要对其进行质量评估，从而选择出最佳的组装结果。质量评估的指标包括基因组的N50、总覆盖度、完整度以及污染度等。N50是指将基因组组装得到的所有叠连群按照长度从大到小进行排序，对排序的叠连群长度依次相加，当相加的总长度达到所有叠连群总长度的一半时，最后相加的叠连群的长度叫作N50。N50主要用来评估基因组组装的质量，其值越大，通常说明组装得到的长片段比例越高。对于基因组的完整度，可通过BUSCO、CheckM等软件进行评价，进而判断基因组组装质量。其中，BUSCO软件的原理是利用直系同源数据库构建出各系统进化分枝的保守基因集，并选择被评估物种所属系统进化分枝相应的保守基因集，根据该基因集比对到新组装的基因组的比例对其完整度进行评估，如图 7-2-2 所示。

After completing the assembly of the genome, it is necessary to evaluate the quality of the assembled genome and choose the best result. The criteria for the assessment include N50, total coverage, genome integrity and contamination of the assembled genome. Only N50 is going to be discussed here as an example. N50 indicates the quality of genome assembly. Generally, the larger the value, the higher the proportion of long fragments assembled. This is accomplished by arranging all contigs obtained in order of the length, and add up the arranged contigs from the longest one until reaching half of the total length of all contigs. The length of the last-added contig is called N50. Software such as BUSCO and CheckM can be used to access the integrity of the assembled genome to further assess the quality of the assembly. BUSCO uses orthologous databases to construct conserved gene sets for different phylogenetic branches. Then the conserved gene sets from the branch to which the assessed species belong to are aligned to the assembled genome, and the aligned proportion can be used for evaluation of genome integrity, as shown in Fig. 7-2-2.

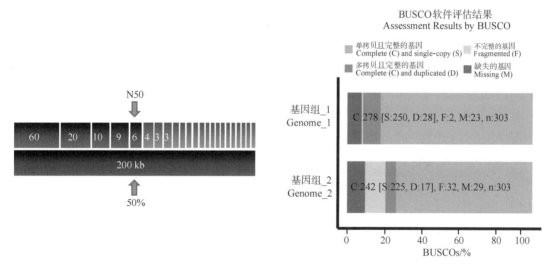

图 7-2-2　BUSCO 软件评估结果

Fig. 7-2-2　BUSCO assessment results

D. 基因组注释

D. Genome annotation

完成基因组组装和质量评估后，可初步得到新物种的基因组。但是未经过基因注释时，该基因组仅包含不同长度的叠连群，每条叠连群中的基因和调控元件相关信息仍然未知。基因组注释就是为了识别、定位和鉴定基因组中每个基因元件，提供具体基因元件的名称、在基因组中的方向以及其起始和终止位置等信息。基因注释过程包括基因预测与基因功能鉴定。

The process of gene annotation is to recognize, map and identify each gene element in the genome, providing the name, the orientation in the genome, and the starting and ending positions of the specific gene element. The process of gene annotation includes gene prediction and gene function identification. To do this, a preliminary genome must first be generated following genome assembly and quality assessment. The preliminary genome is then ready for the next step of gene annotation. Before gene annotation, the preliminary genome contains only the assembled contigs with different lengths, thus, the genes and regulatory elements contained in each contig remain unknown.

对于数据库中没有记录以及首次测序的物种，基因预测主要基于从头计算策略，它依据基因组序列的模式识别来确定基因组功能元件，其中基因组序列的模式识别可基于单基因组或多基因组的比较基因组方法。此外，还有部分软件是通过隐马尔可夫模型的方法来预测基因，如 HMMER。完成基因预测后，将结果比对至不同数据库来鉴定其生物学功能。常用的数据库有 NCBI 蛋白数据库、Pfam 数据库以及 KEGG 数据库。

For species without a draft genome that is already in databases or being sequenced for the first time, gene annotation mainly employs the ab initio calculation strategy, which is based on the pattern recognition of the genome sequences to determine the functional elements in the genome. This pattern recognition of genome sequences relies on the comparative genomic method of single-genome or multiple-genomes. Aside from the above-mentioned method, other software uses hidden Markov model methods to predict genes, such as HMMER. Upon gene prediction completion, the predicted genes can then be compared with different databases to identify their biological functions. NCBI protein, Pfam, and KEGG are the main databases for functional annotation of genes.

E. 进化分析

E. Evolutionary analysis

将从头测序获得的物种基因组和亲缘关系较近的物种基因组一起构建系统发育树，可了解到新测序物种与其他物种的亲缘关系以及进化方向。建树过程可以采用单基因或多基因串联两种不同方法。单基因建树方法简单便捷，其基本原理是通过在不同基因组注释结果中提取出不同物种之间的标志基因，如核糖体 RNA 基因，然后利用 MEGA 软件对提取的序列进行多序列比对并建树。建树的方法主要有四种：邻接法（NJ）、最大似然法（ML）、最大简约法（MP）和贝叶斯法（BI）。这四种方法各有优劣，对于相同的数据集，可以采用两种以上的方法进行建树分析，互相验证。将构建好的树图结果导出为 NWK 格式后，可以通过 iTOL 等网页工具对系统发育树的颜色、线条、背景等进行修饰。

By constructing a phylogenetic tree using the species' genome obtained by de novo sequencing and the genomes of closely-related species, it is possible to understand interspecies relationship and the direction of evolution. There are two different strategies on which constructing phylogenetic trees is based: single gene or multiple genes in tandem. Single-gene based construction is simple and convenient. Marker genes from different species, such as the ribosomal RNA genes, are firstly extracted from genome annotation results, and then compare the multiple sequence alignment with MEGA software, and construct the phylogenetic trees. There are four commonly used methods to construct phylogenetic trees, including neighbor-joining (NJ), maximum likelihood (ML), maximum parsimony (MP), and Bayesian inference (BI). Different methods are equipped with distinct advantages but are also hindered with disadvantages, therefore two or more methods can, and should be used simultaneously for analysis and mutual verification for the same data set to construct phylogenetic trees. Results of phylogenetic tree can then be exported to the files in NWK format, the color, line, background and other formats of the phylogenetic tree can be modified through iTOL or other web-based tools.

多基因串联建树方法的基本原理是对不同物种间的直系同源基因簇使用MAFFT等软件进行多序列比对，然后利用Gblocks等软件将比对后的同一物种内的不同基因按照设定顺序串联起来，最后利用MEGA或RAxML等软件对串联序列建树。相比于单基因构建进化树方法，多基因串联构建的系统发育树在不同物种间的密码子使用偏好性、核苷酸差异率等方面更加拟合物种的实际进化模型。在计算条件允许的情况下，利用同源基因簇构建的系统发育树更符合物种的真实进化历程。

On a different note, when the 2nd strategy, multiple genes in tandem, is used for constructing phylogenetic trees, software such as MAFFT is used to perform multiple sequence alignment of different orthologous gene clusters among different species, and downstream software such as Gblocks is used to concatenate the different genes in a stated order in the same species. Finally, a phylogenetic tree is constructed by using software such as MEGA and RAxML on the basis of the sequences aligned and concatenated. The phylogenetic tree constructed by the 2nd method is more inclined to reflect the true evolutionary model of the species in terms of codon usage preference and the nucleotide variation rate than that constructed by the 1st method. As a result, when computing conditions are not limited, the phylogenetic tree constructed by using orthologous gene clusters correlates with the true evolutionary course of the species than that constructed by using single gene.

2.2 全基因组重测序的数据处理
2.2 Data Processing of Whole-genome Resequencing

全基因组重测序是对已经公开报道过参考基因组的物种进行基因组测序，并在此基础上分析个体或群体差异性，在全基因组水平上检测与表型关联的高频、低频，以及罕见的单核苷酸多态性位点（SNPs）、拷贝数变异（CNV）、插入/缺失（Indel）等变异。图7-2-3所示为全基因组测试分析流程。

Whole-genome resequencing is to sequence the genomes of species that have already been publicly reported reference genomes. Based on such previous work, whole genome resequencing analyzes the differences of individuals or groups, detects high-frequency, low-frequency, and rare single nucleotide polymorphisms (SNPs), copy number variation (CNV), insertion/deletion (Indel) and other mutations associated with phenotypes at the whole genome level. Fig. 7-2-3 shows the whole-genome sequencing analysis pipeline.

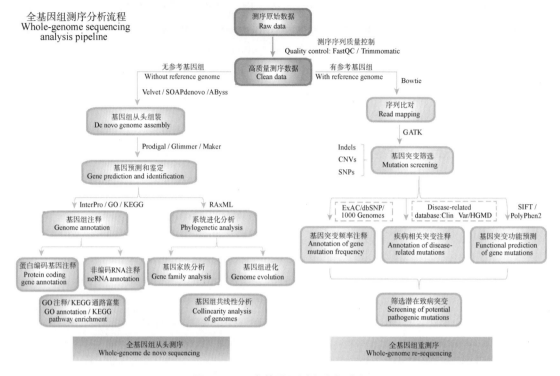

图 7-2-3　全基因组测序分析流程

Fig. 7-2-3　Whole-genome sequencing analysis pipeline

2.2.1　基因组比对
2.2.1　Genome Alignment

基因组重测序分析需要对测序结果进行质量控制，并将过滤后的高质量序列比对到参考基因组上，目前常用的比对软件有 Bowtie、BWA 等，这些软件使用基于 Burrows-Wheeler 变换（BWT）的方法，加快了运行速度，降低了比对过程中的内存使用率。比对过程通常分为两步：第一步使用 Burrows-Wheeler 变换对参考基因组构建索引；第二步将测序序列与参考基因组的索引进行比较，由于参考序列索引被分割成大量的碎片，仅用少量搜索便能识别测序序列的精确位置。比对完成后，比对软件按照 SAM 格式生成序列比对的标准结果。为了节省空间，SAM 文件可由 SAMtools 工具转换成二进制版本的 BAM 文件。

Genome resequencing analysis begins with the quality control of sequencing results, and then alignment of filtered high-quality sequences to the reference genome. Currently, commonly used software for alignment includes Bowtie, BWA, and SOAP2, which are based on the Burrows-Wheeler transformation (BWT) method, which speeds up the analysis and occupies less memory. The alignment process is usually divided into two steps. The 1st step is to use the Burrows-Wheeler transformation to build an index to the reference genome. The 2nd step is to align the reads with said index. The reference sequence is divided and indexed into a large quantity of small segments. The exact location of a given sequence on the reference genome would be identified in just a few searches. After the alignment, the software generates the standard result of the sequence alignment in SAM format. The SAM file can be converted into its binary version by SAMtools to save storage.

SAM 文件主要由头文件和比对结果两部分组成。头文件包含了若干行以@起始的注释信息，分别记录了比对程序使用的参数和参考序列信息、序列 ID 及长度等。比对结果部分是每条序列与参考基因组的比对结果，每条测序读段结果只占一行，每一行分成 12 列，分别记录了不同的信息。每一列代表的详细信息如表 7-2-1 所示，其中第二列的 FLAG 描述序列比对的模式、方向等信息。不同数字代表不同含义，每个 FLAG 代表的含义如表 7-2-2 所示。第六列 CIGAR 表示测序读段比对的具体结果，比如 M 表示能比对上的碱基数，I 和 D 分别表示该位置插入或缺失的碱基数，N 表示跳过这段区域。例如："17S40M25D50M32I" 表示前 17 个碱基比对到其他位置，随后 40 个碱基与参考序列匹配，接下来有 25 个碱基缺失，随后有 50 个碱基匹配，最后有 32 个碱基插入。

The SAM file is typically composed of two parts, the header file and the alignment result. The header file consists of several rows starting with "@", which record the following parameters: the information about the reference genome, sequence ID and length, and other parameters used in the alignment process. The alignment result includes the outcome comparing each read with the reference genome. The outcome from each read occupies only one row, and each row is divided into 12 columns, each of which records different information. The FLAG in the 2nd column describes the mode and direction aligned. Different numbers there represent different meanings. The meaning of the number in the FLAG is shown in Table 7-2-1. The sixth column CIGAR indicates the detailed results of the read alignment. It can be learned from the CIGAR string that M indicates the number of bases that can be aligned, I and D indicate the number of bases inserted into or deleted from the reference respectively, N indicates skipped regions. For example, 17S40M25D50M32I indicates the 1st 17 bases are aligned to other positions, then 40 bases are matched with the reference sequence, the next 25 bases are deleted from it, the following 50 bases are matched, and final 32 bases are inserted into it. For more detailed information, see Table 7-2-2.

表 7-2-1　SAM 文件比对结果中每一列代表的详细信息
Table 7-2-1　The recorded information in each column of SAM alignment result

列 Column	标题 Field	类型 Type	简要描述 Brief description
1 1	QNAME QNAME	字符串 String	测序读段名称 Name of the sequencing read
2 2	FLAG FLAG	整型 Int	FLAG 位标识 Bitwise FLAG
3 3	RNAME RNAME	字符串 String	测序读段比对到的参考基因组名称 The name of the reference genome

续表

列 Column	标题 Field	类型 Type	简要描述 Brief description
4	POS	整型 Int	测序读段比对到参考基因组上的位置 The position of read mapped to the reference genome
5	MAPQ	整型 Int	测序序列比对的质量 Mapping quality
6	CIGAR	字符串 String	简要比对信息（描述了具体的插入、缺失等信息） CIGAR string
7	RNEXT	字符串 String	双端测序的另一条读段比对到的参考基因组名称 Reference genome name of the paired-end read mapped
8	PNEXT	整型 Int	双端测序中另一条读段比对到参考基因组的位置 Position of the paired-end read mapped to the reference
9	TLEN	整型 Int	测序读段长度 The length of the sequencing read
10	SEQ	字符串 String	测序读段的碱基组成 Composition of the sequencing read
11	QUAL	字符串 String	测序读段的质量值 Sequencing quality of the read

表 7-2-2　SAM 文件 FLAG 标签中数字的含义

Table 7-2-2　The meaning of the number in FLAG field

位 Bit	简要描述 Description
1	该测序读段源于双末端测序 The sequencing read is paired
2	双端测序的两条读段被完美比对到参考序列 Each read were properly aligned according to the aligner
4	该读段没有被比对到参考序列 The sequencing read is unmapped
8	双末端测序中该读段对应的另一条读段没有被比对到参考序列 The paired-end read in the template is unmapped
16	该读段被反向互补比对到参考序列 SEQ being reverse complemented
32	双末端测序中该读段对应的另一条读段被反向互补比对到参考序列 SEQ of the paired-end read in the template being reverse complemented

续表

位 Bit	简要描述 Description
64	该读段属于双末端测序的读段 1 The first read in the template
128	该读段属于双末端测序的读段 2 The last read in the template
256	该读段被比对到参考序列的其他位置 Secondary alignment
512	该读段的比对质量不合格 Not passing filters
1 024	该读段源于 PCR 重复 PCR or optical duplicate
2 048	该读段的部分区域被比对到参考序列 Supplementary alignment

2.2.2 基因组突变识别
2.2.2 The Identification of Genome Mutations

基因组突变识别是从测序比对结果中检测出所有与参考序列不同的变异位点，包括单核苷酸变异（SNV）、插入/缺失变异（Indel）、拷贝数变异（CNV）和基因重排等。目前常用的突变识别软件有 GATK 和 Varscan。基因突变识别过程如下。

The mutation identification process is to detect all the mutation sites that are different from the reference sequence in the sequencing alignment results. These include single nucleotide variation (SNV), insertion or deletion (Indel) mutation, copy number variation (CNV) and gene rearrangement, etc. Commonly used mutation identifying software include GATK and Varscan. The process of genome mutation identification is as follows.

完成基因组比对工作后，先通过 SAMtools 工具将 SAM 文件转化为 BAM 二进制文件，随后根据测序读段名称或参考基因组位置对 BAM 文件进行排序，通常 SAMtools 默认按照染色体位置排序。随后使用 Picard 软件去除 PCR 建库过程中产生的具有完全相同序列的 PCR 重复读段。

As mentioned before in the genome alignment excerpt, the SAM file is converted into a BAM binary file by SAMtools after genome alignment, and then the BAM file is sorted according to the read names or reference genome locations. Usually, SAMtools defaults to sort by chromosomal position. Then Picard software is used to remove PCR duplicates. This is necessary because PCR duplicates have the exact same sequences and are generated during PCR in sequencing library preparation.

然而，另一个问题出现了。在比对结果中，Indel 附近会出现大量的碱基错配，这些错配的碱基会被误认为是单核苷酸多态性（SNP），因此需要利用 GATK 对存在潜在序列插入或者序列缺失的区域重新进行比对校正,提高准确性。在完成 Indel 局部重比对后,还需要重新校正碱基质量值(BQSR),该过程主要是通过统计学方法构建测序碱基错误率模型,然后对碱基质量值进行相应调整,消除测序误差,尽量还原物种真实的基因组中变异位点分布,生成新的 BAM 文件。

Yet another problem poses itself. A large number of base mismatches usually exist in the alignment near Indel, which would probably be mistaken as single nucleotide polymorphism (SNP). Therefore, GATK software is used for recalibrating the alignment in the regions with potential insertion or deletion to improve the accuracy of alignment results. After the partial realignment of Indel regions, it is necessary to construct the error rate model of sequencing bases by statistical methods, and then to adjust the quality scores of these bases to eliminate sequencing errors and restore true distribution of mutation sites in the genome of the species to the best possible extent. This is the process of base quality score recalibration (BQSR), which at the end, generates a new BAM file.

最后通过 GATK HaplotypeCaller 模块、MuTect、CNVkit、Lumpy 软件识别 SNV、Indel、CNV 突变的位置,生成存储突变信息的 VCF 报告文件。VCF 文件由两部分组成,第一部分为说明文件,各行均以"##"符号开头。第二部分为突变信息,每行表示一种突变,每一行的信息分为十列来对突变进行详细说明,包括突变在参考基因组中的位置、变异位点的质量值、突变在 dbSNP 数据库中的编号等。

Finally, using software including GATK HaplotypeCaller module, MuTect, CNVkit, or Lumpy, the locations of SNV, Indel, CNV and other mutations are identified, and a VCF (Variant Call Format) file containing information on genome mutations mentioned above is generated. The VCF file consists of two parts. The first part is the description file. Each line starts with two hashtag signs "##". The second part is mutation information, each line represents a mutation and consists of ten columns to show detailed description of the mutation, including its location in the reference genome, the quality score of the mutation site, the ID of the mutation in dbSNP database, etc.

2.2.3 基因组突变过滤
2.2.3 Filtering Genome Mutations

人类全基因组测序数据中通常可以检测到数以百万计的突变,在这些突变中往往有很多是无意义或假阳性的,因此需要对发现的突变进行过滤。突变过滤过程主要分为两部分:首先需要根据各类筛选条件如测序深度、碱基质量等信息进行筛选;随后根据分析对象选择相应过滤用数据库对突变进行过滤,其目的在于过滤掉正常群体中常见的基因组变异,有利于后续筛选出真正和表型相关的有意义突变。目前,常用的突变频率过滤数据库有千人基因组计划数据库、基因组结构变异数据库和人类外显子数据库等。

Millions of mutations usually can be obtained from human genome sequencing. Many of these mutations are often meaningless or false-positive mutations. Therefore, it is necessary to filter mutations discovered from raw human genome sequencing. Two steps make up this process. The first step is choosing various filtering thresholds, such as sequencing depth, base quality and other information. The second step is to use the corresponding filtering databases to filter the mutations according to the analysis object. The aim of the second step is to remove normal genetic variations, benefiting the later identification of meaningful mutations related to specific phenotypes. Currently, the commonly used databases for mutation frequency filtering are from the 1,000 Genomes Project, the Database of Genomic Structural Variation (dbVar) and the Exome Aggregation Consortium (ExAC).

千人基因组数据库是 2008 年由英国 Sanger 研究所、美国国立人类基因组研究所和华大基因研究所（BGI 华大）共同启动的计划，其目标是确定大多数人群中发生频率至少为 1% 的遗传变异，该计划的样本来自 26 个人类种族超过 2 500 名无相关表型信息的个体。千人基因组数据库包含的基因组变异通常来自健康人群，该数据库中出现频率较高的变异通常与病理表型无关，因此通过该数据库可以过滤掉正常人群中常见的变异。ExAC 数据库是目前最大的人类外显子数据库，包含上千万个 DNA 突变，且很多是罕见变异，因此对于罕见遗传病的临床研究和诊断具有重要意义。通过参考 ExAC 数据库中的基因突变频率，研究人员可以排除常见变异，从而迅速锁定真正致病的突变，进而开展相关研究工作。这些基因组数据库，不仅加速了疾病易感基因的发现，还加深了对人类基因组结构差异的认识，为解释人类重大疾病的发病机制、开展疾病个性化预测、预防和治疗奠定了基础。

The 1,000 Genomes Project is jointly initiated by the Sanger Institute in the United Kingdom, the National Human Genome Institute and the Beijing Genomics Institute (BGI) in 2008. The goal is to identify genetic variants with a frequency of at least 1% in most populations. Samples collected are from 2,500 individuals with 26 ethnic backgrounds and no relevant prototypical information. Genetic variations mapped in the 1,000 Genomes Project usually come from healthy people. Variations that appear more frequently in this database tend to be less pathogenic in terms of phenotypes. Therefore, it is possible to utilize this phenomenon to filter out the normal human genome variations using this database. ExAC database is currently the largest human exon database containing tens of millions of DNA mutations, and many of them are rare variations, so it is of great significance for clinical research and diagnosis of rare genetic diseases. Researchers can exclude common variations by referring to the genetic mutation frequency in the ExAC database, so as to quickly identify uncommon mutations that cause diseases and carry out relevant research. These genome databases will not only accelerate the discovery of genetic mutations underlying susceptibility to human diseases, but also deepen the understanding of differences in human genome structure, laying a foundation for the interpretation of the pathogenesis of major human diseases and the development of personalized disease prediction, prevention and treatment.

2.2.4 基因组突变注释与危害程度预测
2.2.4 The Annotation of Genome Mutations and the Prediction of Their Consequences

基因组突变注释是使用各种与疾病关联的突变信息数据库对检测出的突变进行注释的过程。注释的信息主要包括突变发生的基因组坐标、疾病关联信息、参考文献。常用的突变注释数据库包括 HGMD、ClinVar、COSMIC。HGMD 数据库主要从经过同行审议的高质量杂志中搜集并整理与人类遗传疾病密切相关的致病位点，是解析遗传病的重要数据库。HGMD 按照突变与疾病的关联程度，分为致病突变、有待进一步验证的致病突变、有功能性证据且与疾病关联性的突变、有疾病关联性但缺功能性证据的突变、影响基因产物或功能但缺疾病关联性报道的突变。

The genome mutations can be annotated by consulting disease-related mutation databases. The annotation mainly includes the genome coordinates of where the mutations occur, the association relationship with diseases, reference literature. Commonly-used databases include HGMD, ClinVar, COSMIC. An important database for analyzing genetic diseases, the HGMD database, mainly collects and organizes pathogenic sites closely related to human genetic diseases from high-quality journals that have undergone peer review. According to the relations between mutations and diseases, HGMD is categorized into pathogenic mutations, pathogenic mutations that need further verification, disease-related mutations with functional evidence, disease-related mutations without functional evidence, and mutations affecting gene products or functions but without disease-related reports.

突变危害程度预测是利用不同的生物信息学工具辅助解读序列变异在核苷酸及氨基酸水平上的影响。预测工具主要分为两类：一类可以预测突变是否会破坏蛋白质原有功能或结构；另一类可以预测突变是否会影响转录本剪接。部分新工具能够进一步分析非编码区的突变引起的影响。目前包括 CADD、Sift、PolyPhen2 等多款软件均可对突变造成的潜在影响和损害程度进行打分。其中 Sift 软件可以预测突变对蛋白功能的影响，分数越低，说明损害程度越大。

Therisk prediction of the mutation relies on different bioinformatics tools which analyze the effects of the mutation at the nucleotide and amino acid levels. These tools are mainly divided into two categories, some can predict whether the mutation damages the function or structure of the protein, others can predict whether the mutation affects splicing. New tools can also handle additional non-coding sequences. To date, a variety of software including CADD, Sift, PolyPhen2 and others can score the potential consequence and damage caused by mutations. Among these, the Sift software predicts the impact of the mutation on protein function. The lower the score, the higher the harmfulness.

2.3 宏基因组测序的数据处理
2.3 Data Processing of Metagenomic Sequencing

自然界中微生物群落由多种细菌、真菌和其他单细胞微生物构成，它们不仅在自然环境中发挥重要作用，也与人类健康息息相关。受传统培养技术的限制，目前仍然有大量微生物无法在实验室获得纯培养，它们被统称为微生物暗物质。宏基因组测序方法无需经过微生物纯培养，可以直接对环境中的微生物群落进行基因组测序并组装得到不同菌株的基因组，进一步对其进行菌株水平的基因功能注释、比较基因组和进化分析，可以了解微生物暗物质的生态适应机制、营养互作机制以及新陈代谢功能。图 7-2-4 所示为宏基因测序分析流程。

The microbial community in nature is composed of many different bacteria, fungi and other single-celled microorganisms, not only playing important roles in the natural environment but also being closely related to human health. Traditional culture techniques limit a large number of microorganisms, collectively known as microbial dark matter, from being available for pure culture in the laboratory. However, the mixed genomes of different microorganisms in the environment can be directly sequenced by the metagenomic sequencing method, obtaining each genome through genome assembly, thus making the study of non-culturable microorganisms' genomes possible. By performing the annotation of genes and functions, comparative genome and evolution analysis at the strain level on the assembled genomes, the ecological adaptation mechanisms, nutrient interaction mechanisms and metabolic functions of microbial dark matter can be further studied. Fig. 7-2-4 shows metagenome sequencing analysis pipeline.

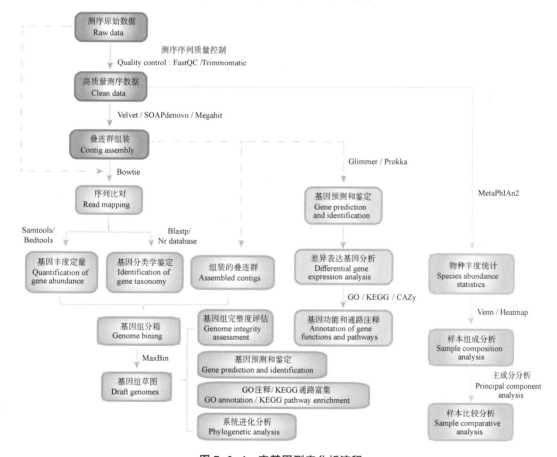

图 7-2-4　宏基因测序分析流程

Fig. 7-2-4　Metagenomie sequencing analysis pipeline

2.3.1　宏基因组分箱
2.3.1　Metagenome Binning

宏基因组分箱是将宏基因组测序读段组装得到的叠连群按物种归类的过程，即将宏基因组数据中来自同一菌株的序列汇集到一起，得到不同菌株的基因组，其方法如下。首先利用基因组组装软件 SPAdes、MEGAHIT 将短的测序读段组装成长的叠连群。然后使用比对软件 Bowtie 将测序序列比对到叠连群上，并通过 SAMtools、bedtools 等软件计算每条叠连群的丰度。接着利用 blast 软件将叠连群比对至 NCBI refseq 蛋白质数据库，获得每条叠连群分类学信息。由于同一个样品中的相同物种应具有相似的叠连群丰度、测序覆盖度、分类学信息以及核苷酸频率，因此 MaxBin 等软件可以根据这些信息对所有叠连群进行基因组分箱，最终得到样本中不同物种的基因组。获得这些不同物种的基因组后，可以通过 CheckM 软件对组装的基因组质量进行评估。CheckM 通常依据编码主要代谢过程的单拷贝基因或者核心保守基因来评估和量化基因组完整度及污染度。后续可以进行基因组预测、功能注释，构建近缘物种参考基因组的系统发育树。

Metagenome binning is a process in which the contigs obtained by the assembly of metagenomic sequencing reads are classified and split by species, the sequences from the same strain in the metagenomic data are grouped together to obtain genomes of different strains, and the details are as follows. Genome assembly software SPAdes, MEGAHIT are first used to assemble short sequencing reads into long contigs. Then, the sequencing reads are aligned to the contigs using alignment software Bowtie, and the abundance of each contig is calculated by SAMtools, bedtools and other software. Next, obtain taxonomic information of each contig by comparing the contigs with NCBI protein refseq data through blast software. Since the same species in a sample should have similar contig abundance, sequencing coverage, taxonomic information, and nucleotide frequency, it is possible for software such as MaxBin to bin all contigs based on this information, and acquire the genomes of different species. Once the genome of the strain is obtained, the quality of the assembled genome can be assessed using CheckM software. CheckM assesses and quantifies genome integrity and contamination based on single-copy genes or core conserved genes encoding major metabolic processes. Subsequently, genome prediction, functional annotation, and phylogenetic analysis of reference genomes of closely related species can be performed.

2.3.2 群落组成分析
2.3.2 The Analysis of Community Composition

微生物群落由成千上万种微生物组成，虽然通过基因组分箱的方法可以获得部分微生物的基因组，但由于仍有部分微生物在环境样品中丰度较低，导致其基因组无法通过分箱的方法获得，因此无法判断环境样品中所有微生物的种类以及相对丰度。为了了解不同样品中微生物的组成和丰度差异，需要使用其他工具进行计算。MetaPhlAn2 是利用宏基因组测序数据对菌群进行定性和定量分析的工具。首先，MetaPhlAn2 软件会根据已有数据库的序列信息形成每个物种独特的标志基因，然后将测序读段与标志基因进行比对，从而确定样品中的物种组成及丰度信息。MetaPhlAn2 目前已经整理了超过 17 000 个物种的参考基因组，包括 13 500 种细菌和古菌、3 500 种病毒和 110 种真核生物，汇编整理了众多物种特异性标志基因。经过 MetaPhlAn2 分析，获得不同种微生物在各样品中的相对丰度矩阵，可以确定各个样品的群落组成与多样性。

The microbial community is composed of thousands of different microorganisms. Some of microbial genomes can be obtained through the method of genome binning, but many low abundance genomes collected from environment samples cannot. Therefore, it is hard to characterize all species and their relative abundance in environmental samples by genome binning. In another attempt to understand the microbial composition and abundance differences in various samples, other tools are used. MetaPhlAn2 is a tool for qualitative and quantitative analysis of microbial communities according to metagenomic sequencing data. First, the MetaPhlAn2 software will form unique marker genes for each species based on the sequence information from existing databases, then compare the sequencing reads with the marker genes to determine the composition and abundance of microorganisms in various samples. MetaPhlAn2 has collected more than 17,000 reference genomes including 3,500 viruses, 13,500 bacteria and archaea, and 110 eukaryotes, sorted out unique marker genes for varied species. Through MetaPhlAn2 analysis, the matrix of relative abundance of different species in various samples is obtained for the downstream determination of community composition and diversity of different samples.

2.3.3 功能基因分析
2.3.3 Functional Gene Analysis

功能基因分析的主要目的是研究宏基因组样品中包含的功能基因，以及这些基因参与的代谢过程，进而分析这些功能基因在微生物群落中发挥的作用。群落功能基因分析有多种不同的方法，例如可以不经过基因组分箱，而是将组装好的叠连群直接通过 MGRAST 等网站进行注释，从宏观上了解该样品中包含哪些代谢过程；或者通过基因预测软件如 Prodigal，用组装好的叠连群预测基因，然后将这些基因分别与 NCBI 数据库、CAZY 数据库、dbCAN 数据库以及 Pfam 数据库进行比对，鉴定其功能，同时也可以利用 KEGG 的 BlastKOALA 功能比对预测好的基因进而完成代谢通路重构，进一步理解微生物群落发挥生物学功能的机制。

Functional gene analysis mainly focuses on the functional genes contained in metagenomic samples and the metabolic processes that these genes participate in. Furthermore, it investigates the functions of these genes in the microbial community. There are a number of methods for functional gene analysis. For example, assembled contigs could be annotated directly through websites such as MGRAST without the process of genome binning to get a macroscopic view of the metabolic processes contained in the sample. Alternatively, gene prediction software such as Prodigal can be used to predict the genes in assembled contigs, and then these genes are aligned to NCBI, CAZY, dbCAN and Pfam databases respectively to identify the functions of these genes. The BlastKOALA function in KEGG can also be used to identify the predicted gene functions and reconstruct the metabolic pathway, extending human understanding of biological mechanisms of the microbial community.

2.4 RNA 测序数据处理
2.4 Data Processing of RNA Sequencing

RNA 高通量测序（RNA-seq）是目前高通量测序技术中应用最广泛的一种技术之一，

可以对不同物种或者特定细胞类型产生的所有转录本进行测序以及表达定量。RNA-seq 可以帮助我们了解正常组织和肿瘤组织之间、药物治疗前后、或不同的发育阶段内不同组织之间各个条件下样本的基因表达差异，从而在转录和转录后水平上探究基因表达和功能，揭示特定生理过程以及疾病发生过程中的分子机理。图 7-2-5 所示为 RNA 测序分析流程。

RNA high-throughput sequencing（RNA-seq）is one of the most widely-used technique in current high-throughput sequencing technology. This technique detects all transcripts and quantifies their expression levels in varied species and specific cell types. RNA-seq can improve current understanding of the differences of gene expression between normal and tumor tissue; before and after drug treatment; or in tissue from various developmental stages. Through these studies reveal gene expression and functions at the transcriptional and post-transcriptional levels, and molecular mechanisms in specific physiological processes and disease development. Fig. 7-2-5 shows RNA sequencing analysis pipeline.

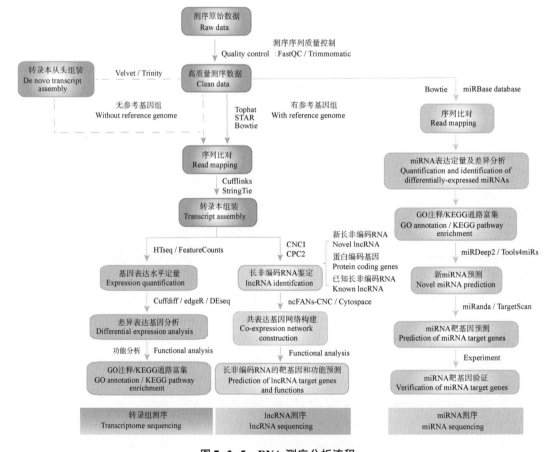

图 7-2-5　RNA 测序分析流程

Fig. 7-2-5　RNA sequencing analysis pipeline

2.4.1 质量控制和基因组比对
2.4.1 Quality Control and Genome Alignment

对 RNA-seq 原始数据进行质量控制的过程如前文所述。获得的高质量测序数据需要通过转录组比对软件 STAR、TopHat 将测序序列比对至参考基因组，比对时所需的参考基因组和 GFF/GTF 注释文件可以从不同网站下载，如 NCBI、UCSC，或者特定物种的核心数据库，如果蝇的 FlyBase。下载参考基因组文件和 GFF/GTF 注释文件时，两个文件版本需保持一致。比如人类参考基因组目前常用的两个版本 GRCh37 和 GRCh38，是 GRC 机构分别于 2009 年和 2013 年发布的。GRCh38 版本的基因组组装比 GRCh37 更加完整，且增加了许多备用基因座（Alternate locus）。备用基因组是指同一处基因组位置上的另一条代表性序列，表示人类基因组多样性。

The RNA-seq data quality control process has been introduced in previous section. Transcriptome alignment software such as STAR and TopHat is required to align the previously obtained high-quality sequencing reads to the reference genome. The required reference genomes and GFF/GTF annotation files can be downloaded from various websites, such as NCBI, UCSC and core databases of specific species like FlyBase for fruit flies. When downloading the reference genome file and GFF/GTF annotation file, make sure to keep the downloaded version of two files consistent. For example, there are currently two commonly used versions of the human genome, GRCh37 and GRCh38, which are the reference sequences of the human genome released by the GRC Institute in 2009 and 2013, respectively. Compared with GRCh37, the GRCh38 version's assembled genome is more complete, with its database having been added more alternate loci. Due to the diversity of the human genome, alternate loci are needed. An alternate locus means a sequence which is an alternate representation of a genomic region.

参考基因组与注释文件准备好后，使用比对软件构建参考基因组索引，随后将测序读段映射到参考基因组上，生成 SAM 格式文件。利用 SAMtools 工具对 SAM 文件进行排序并转化成 BAM 文件，并构建 BAM 文件索引，即可在 IGV 等基因组浏览器对 BAM 文件实现可视化。从基因组浏览器上可以直观地比较实验组与对照组的转录本结构和基因表达量变化。

After downloading the reference genome and the annotation file, the reference genomeis first indexed by the alignment software, and then the sequencing reads are mapped to the reference genome to generate a SAM format file. SAMtools is used to sort the SAM file, and convert it into a BAM file as mentioned before, and further build an index to the BAM file. Now, the generated BAM file can be visualized in genome browsers like IGV. From the genome browser, the changes in transcript structure and gene expression between the experimental and control groups can be compared intuitively.

2.4.2 转录本定量
2.4.2 The Quantification of Transcripts

由于一个基因可以产生多个转录本，RNA-seq 分析的准确性依赖于基因转录异构体（isoform）的重建。因此在进行基因表达定量之前，需要组装转录本，常用软件包括 Cufflinks、StringTie。经过组装会产生新的注释文件，然后将测序序列对应到各转录本上，以统计转录本的表达丰度。表达定量最简单的方法是统计比对到每个转录本上或每个基因上的读段数量，目前有多款软件可以用于转录本定量，如 HTSeq、FeatureCounts、DEXSeq。最终不同样本的定量结果被合并为一个表达矩阵，每一行代表一条转录本或基因，对应的数值为该转录本或基因在不同样本中的表达丰度。

A gene can produce several transcripts, making the accurate analysis of RNA-seq depend on the reconstruction of transcribed gene isoforms. Therefore, the assembly of transcripts is required before the quantification of gene expression. The commonly-used software for transcript assembly includes Cufflinks and StringTie. To quantify transcripts, new annotation files are generated from the assembly, then the sequencing reads are assigned to transcripts and data regarding abundance of transcripts are acquired. The simplest method of expression quantification is to count the number of reads assigned to each transcript or each gene. The software available for the quantification includes HTSeq, FeatureCounts, and DEXSeq. Finally, the quantification results of different samples are combined into an expression matrix. Each row of the expression matrix represents a transcript/gene, and the corresponding value is the abundance of the transcript/gene in the samples.

2.4.3 归一化
2.4.3 Normalization

在 RNA-seq 分析中，测序深度以及转录本长度是影响基因表达定量的两个重要因素。转录本越长，测序深度越大，比对到该转录本的测序读段数就会越多，从而导致无法对基因表达进行准确定量。因此，需要对不同样本中所有的转录本进行归一化处理。目前主要有三种对基因表达进行定量的参数：RPKM，即每一千碱基转录本上每百万条总比对读段中检测到的该转录本读段数；FPKM，即每一千碱基转录本上每百万条总比对读段中检测到的该转录本成对读段数；TPM，即每百万条总比对读段中检测到的该转录本数。基因表达定量参数的选择要根据 RNA-seq 是单末端还是双末端测序，以及标准化处理测序深度和转录本长度的先后顺序来决定。

All transcripts in different samples need to be normalized. This is because, in RNA-seq analysis, sequencing depth and transcript length are two important factors that affect the quantification of gene expression. The longer the transcript, and the higher the sequencing depth, the more sequencing reads will be mapped to the transcript, thus making it difficult to accurately quantify gene expression. There are currently three main parameters for quantifying gene expression. RPKM: reads per Kilobase of transcript per million mapped reads; FPKM: fragments per kilobase of transcript per million mapped reads; and TPM: transcripts per million mapped reads. The selection of parameters for quantifying gene expression depends on whether RNA-seq is single-end or paired-end sequencing, and the order in which sequencing depth and transcript length standardization occurs.

2.4.4 差异表达分析
2.4.4 Differential Expression Analysis

差异表达分析是通过统计学方法鉴定不同条件样品中表达差异显著的基因，进而分析这些基因参与生物过程的机制。目前常用的差异表达分析软件有 edgeR、DESeq。在进行差异表达分析前，为了减少计算量，通常先从表达矩阵过滤掉所有样品里一贯低表达的转录本。再利用 edgeR、DESeq 对表达矩阵中不同条件进行归一化处理，并计算两个条件之间基因差异表达倍数的对数值（$\log_2$FoldChange）和校正后的显著性 p 值（p-value，probability），筛选出表达差异显著的基因。

The purpose of differential expression analysis is to identify genes with significant expression differences under varied conditions, done so through statistical methods, and further analyze the mechanisms of how these genes are involved in biological processes. Currently, commonly used software of differential expression analysis includes edgeR and DESeq. For the purpose of reducing the amount of calculation, the transcripts with consistent low expression in all samples are usually filtered out from the expression matrix before differential expression analysis. Then using edgeR or DESeq, the expression levels under different conditions inside the matrix are normalized, $\log_2$FoldChange (logarithm of expression fold change) and corrected p-value (probability, significance of the difference) between two conditions are calculated, and differentially expressed genes (DEGs) are identified according to these two parameters.

2.4.5 功能富集分析
2.4.5 Functional Enrichment Analysis

差异表达基因功能富集分析主要将显著差异表达的基因按照功能进行分类，进一步与生物学表型进行关联。Gene ontology（GO）和 kyoto encyclopedia of Genes and Genomes（KEGG）是两种常见的基因功能注释数据库，可以将基因按照功能进行分类。

The functional enrichment analysis with differentially expressed genes is to classify these genes according to their functions, and further correlate the functions with the biological phenotypes. Gene ontology (GO) and kyoto encyclopedia of genes and genomes (KEGG) are two databases that annotate gene functions, meaning the above databases can classify genes according to their functions.

GO 注释提供了一系列的术语条目（terms）用于描述基因、基因产物的特性。这些条目可以归为三类：cellular component（CC）用于描述基因的作用位置，即基因产物所在的亚细胞结构、位置和所属的大分子复合物信息；molecular function（MF）用于描述基因产物所富集的功能；biological process（BP）用于描述基因参与的生物学过程，如有丝分裂或嘌呤代谢等。KEGG 是一个整合了基因组信息、化学信息、系统信息及疾病信息的综合数据库，能够利用图形来展现差异表达基因参与的代谢途径、信号传导与基因表达调控通路，以及各途径之间的关系。

GO annotation provides a series of terms describing the characteristics of genes and gene products. These terms are divided into three distinct categories: Cellular component (CC), describes subcellular locations in which genes play functions, providing information on subcellular structures, loci, and macromolecular complexes where the gene products are located; Molecular function specifies enriched molecular roles of gene products; Biological process defines the biological processes that the genes are involved in, such as mitosis and purine metabolism. KEGG is a comprehensive database, integrating genomic, chemical, systemic information and disease information. KEGG is able to use graphics to visualize the metabolic pathways, signaling transduction pathways, and gene expression control pathways, as well as the relationship among the pathways in which differentially expressed genes take part.

功能富集分析可以在 Metascape、DAVID 等网站上完成，将基因列表输入这些网站并选择相应物种及功能分析选项后，网站可以自动统计这些基因被显著富集到的 GO 条目和 KEGG 代谢通路。

Functional enrichment analysis canbe completed on the websites such as Metascape and DAVID. To execute such an analysis, the list of gene symbols is entered onto the website and the corresponding species and the type of functional analysis are chosen. The websites will subsequently and automatically compute the GO terms and KEGG metabolic pathways that genes are enriched in.

参考文献

［1］ANTONOVSKY N, GLEIZER S, MILO R. Engineering carbon fixation in E. coli: from heterologous RuBisCO expression to the Calvin-Benson-Bassham cycle ［J］. Current Opinion in Biotechnology, 2017, 47: 83-91.

［2］DE SCHOUWER F, CLAES L, VANDEKERKHOVE A, et al. Protein-rich biomass waste as a resource for future biorefineries: state of the art, challenges, and opportunities ［J］. Chem. Sus. Chem. , 2019, 12 (7): 1272-1303.

［3］DELLOMONACO C, RIVERA C, CAMPBELL P, et al. Engineered respiro-fermentative metabolism for the production of biofuels and biochemicals from fatty acid rich feedstocks ［J］. Applied and Environmental Microbiology, 2010, 76: 5067-5078.

［4］HENNIG C, BROSOWSKI A, MAJER S. Sustainable feedstock potential—a limitation for the bio-based economy ［J］. Journal of Cleaner Production, 2016, 123: 200-202.

［5］HUO Y X, MYUNG C K, RIVERA J, et al. Conversion of proteins into biofuels by engineering nitrogen flux ［J］. Nature Biotechnology, 2011, 29: 346-51.

［6］LEE T C, XIONG W, PADDOCK T, et al. Engineered xylose utilization enhances bioproducts productivity in the cyanobacterium synechocystis SP. PCC 6803 ［J］. Metabolic Engineering, 2015, 30: 179-189.

［7］SEKAR R, SHIN H D, DICHRISTINA T. Activation of an otherwise silent xylose metabolic pathway in shewanella oneidensis ［J］. Applied and Environmental Microbiology, 2016, 82 (13): 3996-4005.

［8］SHEN Y. Carbon dioxide bio-fixation and wastewater treatment via algae photochemical synthesis for biofuels production ［J］. RSC Adv. , 2014, 4 (91): 49672-49722.

［9］CALERO P, NIKEL P I, Chasing bacterial chassis for metabolic engineering: aperspective review from classical to non-traditional microorganisms ［J］. Microbial Biotechnology, 2019, 12: 98-124.

［10］CHI H, WANG X, SHAO Y, et al. Engineering and modification of microbial chassis for systems and synthetic biology ［J］. Synthetic and Systems Biotechnology, 2019, 4: 25-33.

［11］NIKEL P I, CHAVARRÍA M, DANCHIN A, et al. From dirt to industrial applications: pseudomonas putida as a synthetic biology chassis for hosting harsh biochemical reactions ［J］. Current Opinion in Chemical Biology, 2016, 34: 20-29.

［12］LI M, HOU F, WU T, et al. Recent advances of metabolic engineering strategies in natural isoprenoid production using cell factories ［J］. Natural Product Reports, 2020, 37 (1): 80-99.

［13］CHAVEZ R, FIERRO F, GARCÍA R, et al. Filamentous fungi from extreme environments as a promising source of novel bioactive secondary metabolites [J]. Frontiers in Microbiology, 2015, 6: 903.

[14] 赵晓祥, 张小凡. 环境微生物技术 [M]. 北京: 中国环境工业出版社, 2015.

[15] CHAKRABORTY R, WU C H, HAZEN T C. Systems biology approach to bioremediation [J]. Current Opinion in Biotechnology, 2012, 23 (3).

[16] 张兰英, 刘娜, 王显胜. 现代环境微生物技术 [M]. 2版. 北京: 清华大学出版社, 2007.

[17] SINGH L R. Principles and applications of environmental biotechnology for a sustainable future biosensors [M]. Singapore: Springer, 2017: 341-363.

[18] NGUYEN V T, KWON Y S, GU M B. Aptamer-based environmental biosensors for small molecule contaminants [J]. Current Opinion in Biotechnology, 2017, 45: 15-23.

[19] TAN L, SCHIRMER K. Cell culture-based biosensing techniques for detecting toxicity in water [J]. Current Opinion in Biotechnology, 2017, 45: 59-68.

[20] MCGHEE C E, LOH K Y, LU Y. DNAzyme sensors for detection of metal ions in the environment and imaging them in living cells [J]. Current Opinion in Biotechnology, 2017, 45: 191-201.

[21] JERNELÖV A. How to defend against future oil spills [J]. Nature, 2010, 466 (7303): 182-183.

[22] RON E Z, ROSENBERG E. Enhanced bioremediation of oil spills in the sea [J]. Curr Opin Biotechnol, 2014, 27 (27C): 191-194.

[23] CRAWFORD R L, ROSENBERG E. The Prokaryotes [J]. 4th ed. New York: Springer, 2012.

[24] 赵远, 梁玉婷. 石油化工环境生物技术 [M]. 北京: 中国石化出版社, 2013.

[25] HAZEN T C, DUBINSKY E A, DESANTIS T Z, et al. Deep-sea oil plume enriches indigenous oil-degrading bacteria [J]. Science, 2010, 330 (6001): 204-208.

[26] WANG W, SHAO Z. The long-chain alkane metabolism network of alcanivorax dieselolei [J]. Nature Communications, 2014, 5 (5755): 1-28.

[27] 王加龙. 废旧塑料回收利用实用技术 [M]. 北京: 化学工业出版社, 2010.

[28] 戈进杰. 生物降解高分子材料及其应用 [M]. 北京: 化学工业出版社, 2002.

[29] RAY S. 生物降解聚合物及其在工农业中的应用 [M]. 北京: 机械工业出版社, 2010.

[30] YANG Y, YANG J. WU W M, et al. Biodegradation and mineralization of polystyrene by plastic-eating mealworms: Part 1. Chemical and Physical Characterization and Isotopic Tests [J]. Environmental Science & Technology, 2015, 49 (20): 12080.

[31] YANG Y, YANG J, WU W M, et al. Biodegradation and mineralization of polystyrene by plastic-eating mealworms: part 2. Role of Gut Microorganisms [J]. Environmental Science & Technology, 2015, 49 (20): 12087-12093.

[32] YOSHIDA S, HIRAGA K, TAKEHANA T, et al. A bacterium that degrades and assimilates poly (ethylene terephthalate) [J]. Science, 2016, 351 (6278): 1196-1199.

［33］毛海龙，白俊岩，姜虎生，等. 可降解塑料的微生物降解研究进展［J］. 微生物学杂志，2014，34（4）：80-84.

［34］李凡，王莎，刘巍峰，等. 聚乳酸（PLA）生物降解的研究进展［J］. 微生物学报，2008，48（2）：262-268.

［35］史可，苏婷婷，王战勇，等. 可降解塑料聚乳酸（PHA）生物降解性能进展［J］. 塑料，2019，48（3）：36-41.

［36］JENNIFER A D, EMMANUELLE C. The new frontier of genome engineering with CRISPR—Cas9［J］. Science, 2014, 346（6213）: 1258096-1.

［37］ALLISON H O, TAE S M. Programmable genetic circuits for pathway engineering［J］. Current Opinion in Biotechnology, 2015, 36: 115-121.

［38］HOLLINGSWORTH S A, DROR R O. Molecular dynamics simulation for all［J］. Neuron, 2018, 99（6）: 1129-1143.

［39］JOSHUA H, GIULIA Z. Fine details in complex environments: the power of cryo-electrontomography［J］. Biochemical Society Transactions, 2018, 46（4）: 807-816.

［40］YIGONG S. A Glimpse of structural biology through X-Ray crystallography［J］. Cell, 2014, 159（5）: 995-1014.

［41］ISABEL M, GWYNDAF E, JUAN S W, et al. Membrane protein structure determination—the next generation［J］. Biochimica Et Biophysica Acta, 2014, 1838（1）: 78-87.

［42］ROUT M P, SALI A. Principles for integrative structural biology studies［J］. Cell, 2019, 177（6）: 1384-1403.

［43］SHI Y G. A Glimpse of structural biology through X-Raycrystallography［J］. Cell, 2014, 159（5）, 995-1014.

［44］STACY A M, DANA C, NADLER R Y, et al. Biofuel metabolic engineering with biosensors［J］. Current Opinion in Chemical Biology, 2016, 35: 150-158.

［45］ZENG W Z, LI G, XU S, et al. High-throughput screening technology in industrial biotechnology. Trends in Biotechnology, 2020, 38（8）: 888-906.

［46］ANTHONY D C, ROXANNA B, MIRIAM L G. The use of fluorescence-activated cell sorting in studying plant development and environmental responses［J］. The International Journal of Developmental Biology, 2013, 57: 545-552.

［47］EDEL M, VÖLLER K, REINHOLZ T, et al. Integrated biomass policy frameworks［R］. London: Intelligent Energy for Europe Program of the European Union, 2017.

［48］KING D. The future of industrialbiorefineries［J］. World Economic Forum, 2010: 1-40.

［49］HOLLINSHEAD W, HE L J TANG Y J. Biofuel production: an odyssey from metabolic engineering to fermentation scale-up［J］. Frontiers in Microbiology, 2014, 5: 344.

［50］RAMANAN S, HYUN D S, THOMAS J, et al. Activation of an otherwise silent xylose metabolic pathway inShewanella oneidensis［J］. Applied and Environmental Microbiology, 2016, 82（13）: 3996-4605

［51］JOSHUA S Y, KELLY H T ILLER, HANI AA, et al. Plants to power: bioenergy to fuel the future [J]. Trends in Plant Science, 2008, 13 (8): 421-429.

[52] HUO Y X, KWANG M C, JIMMY G, et al. Conversion of proteins into biofuels by engineering nitrogenflux [J]. Nature Biotechnology, 2011, 29, 346-351.

[53] FREE D S, LAURENS C, ANNELIES V, et al. Protein-rich biomass waste as a resource for future biorefineries: state of the art, challenges, and opportunities [J]. Chem Sus Chem, 2019, 12 (7): 1272-1303.

[54] CLEMENTINA D, CARLOS R, PAUL C, et al. Engineered respiro-fermentative metabolism for the production of biofuels and biochemicals from fatty acid-rich feedstocks [J]. Applied and Environmental Microbiology, 2010, 76: 5067-5078.

[55] RAM C, SHEELU Y, VINEET K. Microbial degradation of lignocellulosic waste and its metabolicproducts [J]. Environmental Waste Management, 2015.

[56] RACHANA S, PARUL P, MADHULIKA S, et al. Uncovering potential applications of cyanobacteria and algal metabolites in biology, agriculture and medicine: current status and futureprospects [J]. Frontiers in Microbiology, 2017, 8: 515.

[57] NIV A, SHMUEL G, RON M. Engineering carbon fixation in E. coli: from heterologousRuBisCO expression to the Calvin-Benson-Bassham cycle [J]. Current Opinion in Biotechnology, 2017, 47: 83-91.

[58] ATSUMI S, HANAI T, LIAO J C. Non-fermentative pathways for synthesis of branched-chain higher alcohols asbiofuels [J]. Nature, 2008, 451 (7174): 86-89.

[59] LYNDSAY E S, REZA P. Glyphosate in runoff waters and in the root-zone: a review [J]. Toxics, 2015, 3 (4): 462-480.

[60] SUN P, SCHUURINK R C, CAISSARD J C, et al. My way: Noncanonical biosynthesis pathways for plant volatiles [J]. Trends in Plant Science, 2016: 884-894.

[61] AKIMASA M. Structure and function of polyketide biosynthetic enzymes: various strategies for production of structurally diverse polyketides [J]. Bioscience, Biotechnology, and Biochemistry, 2017, 81: 12, 2227-2236.

[62] SUN Z H, BÁLINT F, ALESSANDRA D S, et al. Bright side of lignin depolymerization: toward new platform chemicals [J]. Chemical Reviews, 2018, 118 (2): 614-678.

[63] LI M J, HOU FF, WU T, et al. Recent advances of metabolic engineering strategies in natural isoprenoid production using cell factories [J]. Natural Product Reports, 2020, 37: 80-99.

[64] RIFAT H, SAFDAR A, UMMAY A, et al. Soil beneficial bacteria and their role in plant growth promotion: a review [J]. Annals of Microbiology, 2010, 60 (4): 579-598.

[65] SILVIA P, BRENDAN B J M. Ecology of nitrogen fixing, nitrifying, and denitrifying microorganisms in tropical forestsoils [J]. Frontiers in Microbiology, 2016, 7 (373).

[66] ANAMIKA D, ASHWANI K, ABEER H, et al. Growing more with less: breeding and developing drought resilient soybean to improve food security [J]. Ecological Indicators, 2019, 105: 425-437.

[67] YOON Y J, KIM E S, HWANG Y S, et al. Avermectin: biochemical and molecular

basis of its biosynthesis and regulation [J]. Applied Microbiology and Biotechnology, 2004, 63 (6): 626-634.

[68] GHOLAMREZA S J, ELENA V, REZA S. Bacillus thuringiensis: A successful insecticide with new environmental features andtidings [J]. Applied Microbiology and Biotechnology, 2017, 101 (7): 2691-2711.

[69] ANDRÉ M N, PATRÍCIA A, HUGO C P, et al. Antifungal drugs: new insights in research &development [J]. Pharmacology & Therapeutics 2019, 195, 21-38.

[70] EÓIN O, ANDRÉ L A N, GUAN L L. The role of the gut microbiome in cattle production and health: driver or passenger [J]. Annual Review of Animal Biosciences, 2020, 8: 199-220.

[71] PAULINA M, KATARZYNA Ś. The role of probiotics, prebiotics and synbiotics in animal nutrition [J]. Gut Pathogens, 2018, 10 (1): 21.

[72] ASHIMA V, POONAM S, ANSHU M. Probiotic yeasts in livestock sector [J]. Animal Feed Science and Technology, 2016, 219, 31-47.

[73] ROBIN M, MICHAEL N A, DIMITRIOS T, et al. Gut microbiome metagenomics to understand how xenobiotics impact human health [J]. Current Opinion in Toxicology, 2018, 11-12: 51-58.

[74] CHRISTOPHER Q, ALAN W W, JARED T S, et al. Shotgun metagenomics, from sampling to analysis [J]. Nature Biotechnology, 2017, 35 (9): 833-844.

[75] ZHAO J B, BAI Y, TAO S Y, et al. Fiber-rich foods affected gut bacterial community and short-chain fatty acids production in pig model [J]. Journal of Functional Foods, 2019, 57: 266-274.

[76] HARRY J, FLINT K P, PETRA L, et al. The role of the gut microbiota in nutrition and health [J]. Nature Reviews Gastroenterology & Hepatology 2012, 9 (10): 577.

[77] YI SHANG, YONGSHUO MA, YUAN ZHOU, et al. Biosynthesis, regulation, and domestication of bitterness in cucumber [J]. Science, 2014, 346 (6213): 1084-1088.

[78] BROWN S M. Next-generation DNA sequencing informatics, [M]. 2nd ed. New York: Cold Spring Harbor Laboratory Press, 2015.

[79] CHANG H, LIM J, HA M, et al. Tail-seq: genome-wide determination of poly (A) tail length and 3' end modifications [J]. Mol Cell, 2014, 53 (6): 1044-1052.

[80] KATANNYA K, YEO G W. Genome-wide approaches to dissect the roles of RNA binding proteins in translational control: implications for neurological diseases [J]. Frontiers in Neuroscience, 2012, 6: 144.

[81] KOBOLDT D C, MARDIS E R, STEINBERG K M, et al. The next-generation sequencing revolution and its impact on genomics [J]. Cell, 2013, 155 (1): 27-38.

[82] TUIMALA J. RNA-seq data analysis: A practical approach [J]. CRC Press, 2014.

[83] KORPELAINEN E, TUIMALA J, SOMERVUO P, et al. RNA-seq data analysis: A practical approach: CRC Press; 2014.

[84] MARDIS E R. A decade's perspective on DNA sequencing technology [J].

Nature, 2011.

［85］MARDIS E R. DNA sequencing technologies：2006-2016 ［J］. Nat Protoc 2017, 12 (2)：213-218.

［86］STARK R, GRZELAK M, HADFIELD J. RNA sequencing：the teenage years ［J］. Nat Rev Genet 2019, 20 (11)：631-656.

［87］SUBTELNY A O, EICHHORN S W, CHEN G R, et al. Poly(A) tail profiling reveals an embryonic switch in translational control ［J］. Nature 2014, 508 (7494)：66-71.

［88］李金明. 高通量测序技术 ［M］：北京：科学出版社，2018.

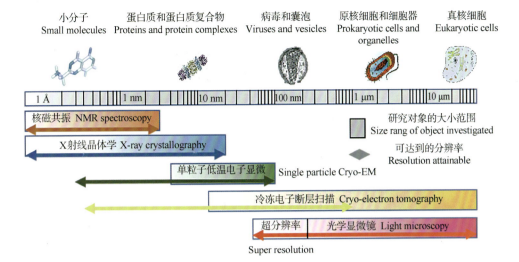

图 1-7-1　结构生物学技术及其研究对象

Fig. 1-7-1　Research objects and tools of structural biology

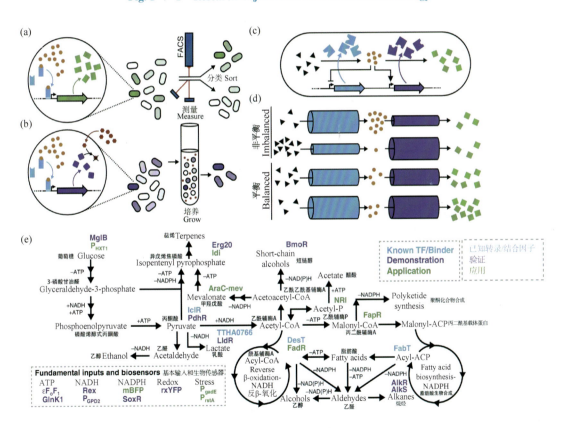

图 1-8-2　生物传感器的应用

Fig. 1-8-2　Applications of biosensors

（a）一个以荧光灯信号作为输出的生物传感器；（b）具有选择性输出的生物传感器；
（c）用于途径动态调控的生物传感器；（d）途径平衡；（e）潜在的生物燃料合成途径和生物传感器

(a) A biosensor with an output；(b) A biosensor with a selectable output；(c) Biosensors for dynamic regulation of pathways；(d) Visualization of pathway balancing；(e) Potential biofuel pathways and biosensors

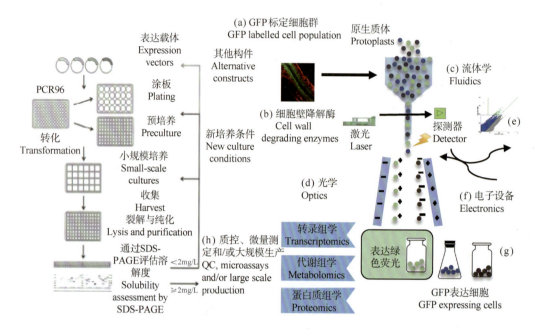

图 1-9-2 荧光激活细胞分选的工作流程

Fig. 1-9-2 Fluorescence-activated cell sorting workflow

（a）选择携带荧光报告基因的转基因植物；（b）用酶处理收获样品并过滤；
（c）在流式细胞仪中处理样本；（d）联合激光和探测仪分析以定义液滴；（e）分析发射光谱；
（f）带有电荷的液滴；（g）所有其他细胞进行废物收集的双向分选；（h）收集细胞并分析

（a）Choose transgenic plants carrying a fluorescent reporter；（b）Sampling harvesting, treating with enzymes and filtering；
（c）Dealing with the sample in a FACS machine；（d）Measure the fluorescence and other properties of each droplet；
（e）Analyzing the emission spectrum；（f）Droplets Charged；（g）Waste collection of all other cells；（h）Cells collecting and analyzing

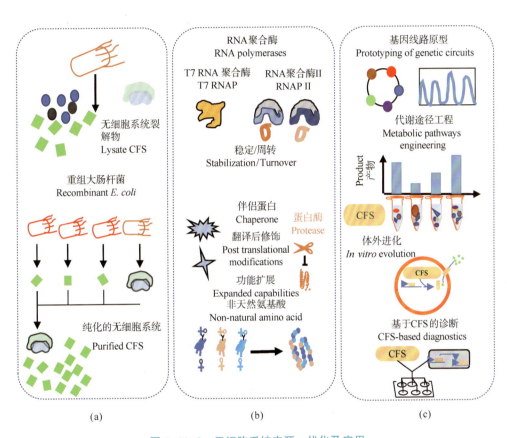

图 1-11-3 无细胞系统来源、优化及应用
Fig. 1-11-3 CFS source, optimization and application

（a）合成生物学方法制备 CFS；（b）CFS 的优化和扩展；（c）CFS 在合成生物学中的应用
(a) CFS for synthetic biology approaches; (b) Optimization and expansion of CFS; (c) Applications for CFS in synthetic biology

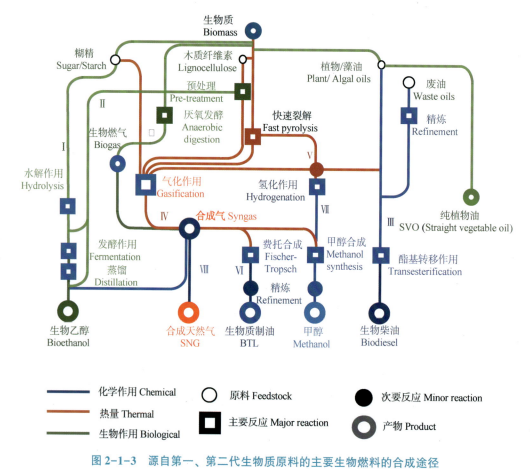

图 2-1-3 源自第一、第二代生物质原料的主要生物燃料的合成途径

Fig. 2-1-3 The synthetic conversion routes of major biofuels produced from the 1st and 2nd generation biomass feedstock

注释：
Ⅰ. 糖/淀粉作物的发酵；Ⅱ. 木质纤维素生物质的发酵；Ⅲ. 甘油三酯的酯交换；Ⅳ. 合成气；Ⅴ. 快速裂解；Ⅵ. 费托合成；Ⅶ. 氢化；Ⅷ. 合成天然气

Note：
Ⅰ. Fermentation of sugar/starch crops；Ⅱ. Fermentation of lignocellosic biomass；Ⅲ. Transesterification of triglycerides；Ⅳ. Syngas；Ⅴ. Fast pyrolysis；Ⅵ. Fischer-Tropsch synthesis；Ⅶ. Hydrogenation；Ⅷ. Synthetic natural gas（SNG）

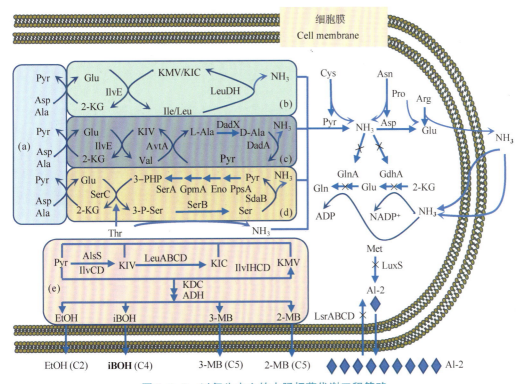

图 2-1-9 以氮为中心的大肠杆菌代谢工程策略

Fig. 2-1-9 Nitrogen-centric metabolic engineering strategy in *E. coli*

(a) 天冬氨酸和丙氨酸中的氨基转移得到丙酮酸和谷氨酸；(b) 谷氨酸氨基转移得到异亮氨酸和亮氨酸；
(c) 谷氨酸氨基转移得到丙酮酸；(d) 谷氨酸中氨基转移到丙酮酸；(e) 工程酮酸途径

(a) Amino groups in Asp and Ala are transferred to Pyr and Glu; (b) Amino groups in Glu are transferred to Ile and Leu;
(c) Amino groups are transferred to Pyr; (d) Amino groups in Glu are transferred to Pyr;
(e) Engineered keto acid pathways.

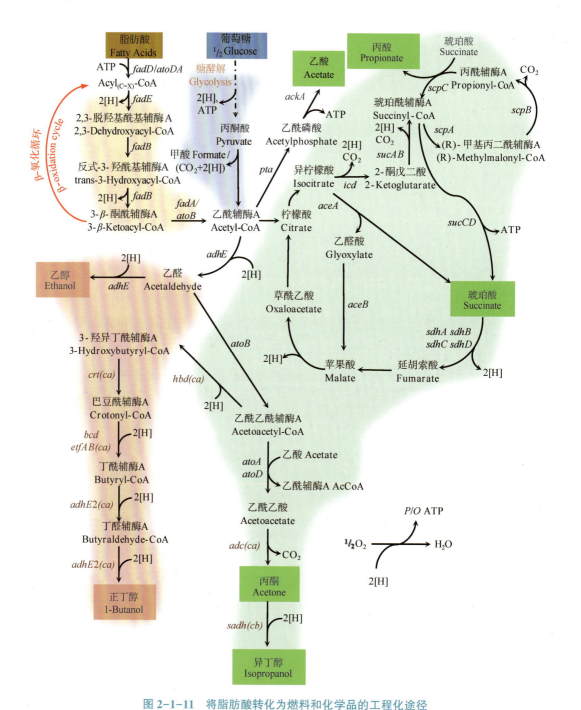

图 2-1-11 将脂肪酸转化为燃料和化学品的工程化途径

Fig. 2-1-11 Pathways engineered in *E. coli* for the conversion of fatty acids to fuels and chemicals

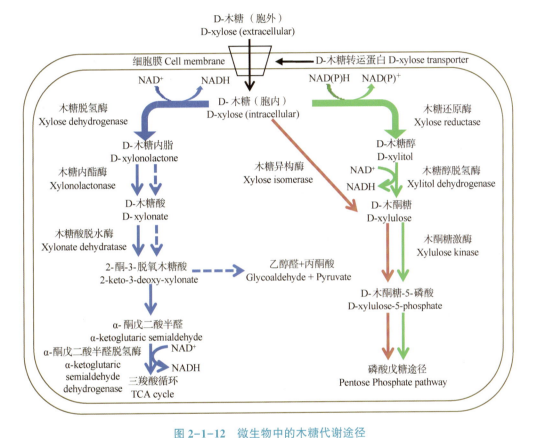

图 2-1-12 微生物中的木糖代谢途径
Fig. 2-1-12 Xylose metabolic pathways in microorganisms

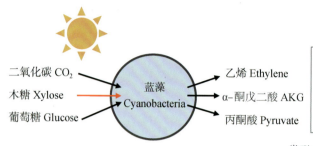

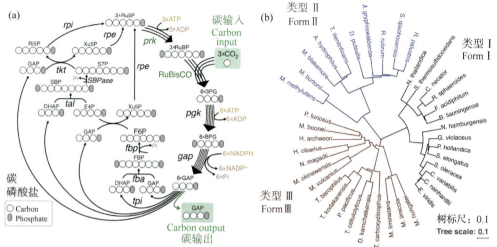

图 2-1-16 木糖利用途径工程化设计

Fig. 2-1-16 Engineered xylose utilization

（a）CBB 途径中的酶和中间体的代谢；（b）RuBisCO 大亚基序列的系统发生树

(a) Metabolic diagram of the enzymes and intermediates in the CBB pathway;

(b) Phylogenetic tree of RuBisCO large subunit sequences

注释：

RuBP，1,5-二磷酸核酮糖；Ru5P，5-磷酸核酮糖；3PG，3-磷酸甘油酸；BPG，1,3-二磷酸甘油酸；GAP，甘油醛 3-磷酸；DHAP，磷酸二羟丙酮；FBP，1,6-二磷酸果糖；F6P，6-磷酸果糖；Xu5P，5-磷酸木酮糖；E4P，4-磷酸赤藓糖；SBP，1,7-二磷酸景天庚酮糖；S7P，7-磷酸景天庚酮糖；Ri5P，5-磷酸核糖；*rpi*：5-磷酸核糖异构酶；*rpe*：磷酸核酮糖差向异构酶；*pgk*：磷酸甘油酸激酶；*gap*：甘油醛-3-磷酸脱氢酶；*tpi*：磷酸丙糖异构酶；*fba*：1,6-二磷酸果糖醛缩酶；*fbp*：果糖 1,6-二磷酸酶；*tal*：转醛醇酶；*SBPase*：景天庚酮糖二磷酸酶；*tkt*：转酮醇酶。

Note：

RuBP, ribulose-1, 5-bisphosphate; Ru5P, ribulose-5-phosphate; 3PG, 3-phosphoglycerate; BPG, 1, 3-bisphosphoglycerate; GAP: glyceraldehyde 3-phosphate; DHAP: dihydroxyacetone phosphate; FBP: fructose 1, 6-bisphosphate; F6P: fructose 6-phosphate; Xu5P: xylose 5-phosphate; E4P: erythritose 4-phosphate; SBP: sedoheptulose-1, 7-bisphosphate; S7P: sedoheptulose-7-phosphate; Ri5P: ribose-5-phosphate; *rpi*: ribose-5-phosphate isomerase; *rpe*: ribulose-phosphate 3-epimerase; *pgk*: phosphoglycerate kinase; *gap*: glyceraldehyde-3-phosphate dehydrogenase; *tpi*: triosephosphate isomerase; *fba*: fructose-1, 6-bisphosphate aldolase; *fbp*: fructose 1, 6-bisphosphatase; *tal*: transaldolase; *SBPase*: sedoheptulose-bisphosphatase; *tkt*: transketolase.

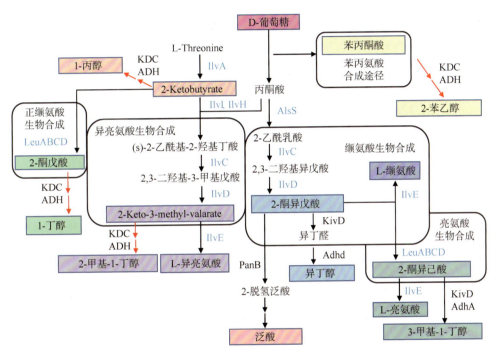

图 2-2-1　工程化大肠杆菌中生产高级醇的合成网络

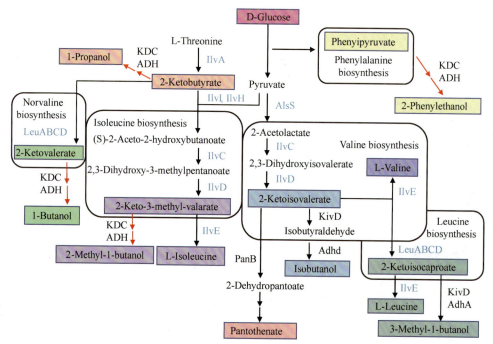

Fig. 2-2-1 The synthetic networks for the higher alcohols production in engineered *E. coli.*

注释：
Note：
IlvA，苏氨酸脱氨酶；IlvC，酮酸还原异构酶；IlvD，二羟基酸脱水酶；IlvH，乙酰乳酸合成酶；IlvE，链氨基酸转移酶；AlsS，乙酰乳酸合成酶；KivD，α-酮异戊酸脱羧酶；Adh，乙醇脱氢酶。

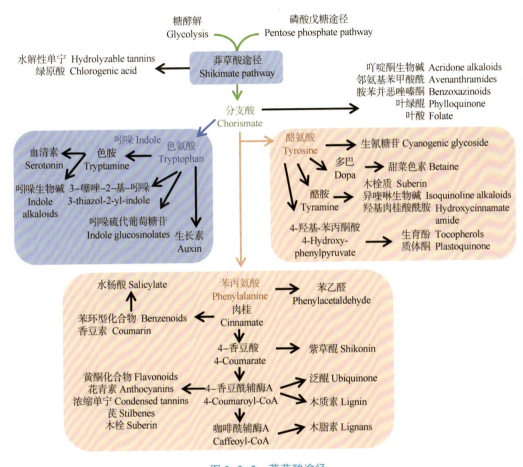

图 2-2-2 莽草酸途径
Fig. 2-2-2　The shikimic acid pathway

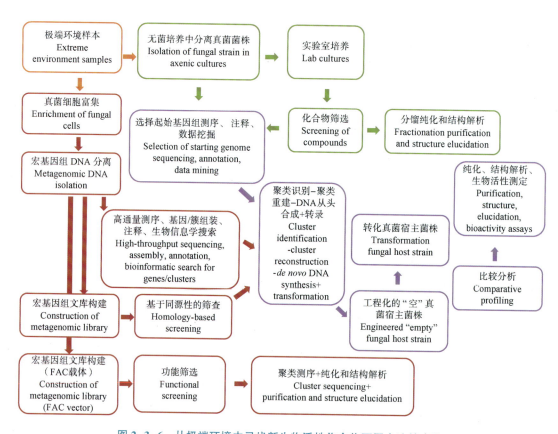

图 2-3-6 从极端环境中寻找新生物活性化合物不同方法的步骤

Fig. 2-3-6 Main steps in different approaches to search for new bioactive compounds from fungi in extreme enrironments

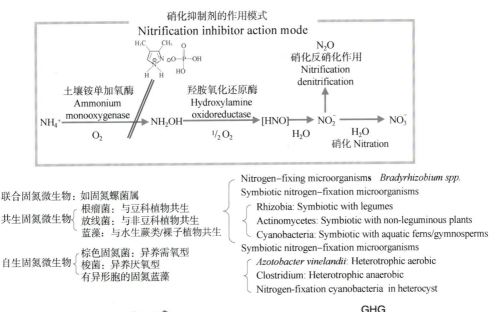

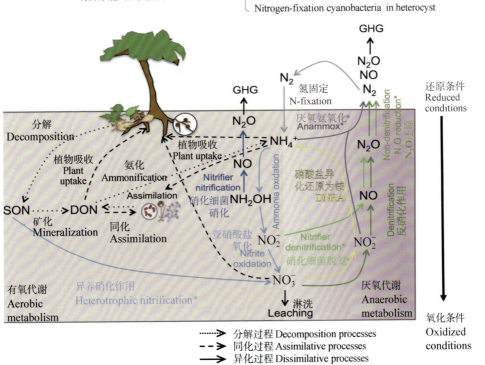

图 3-1-1 热带森林土壤生物氮循环的示意图

Fig. 3-1-1 Schematic representation of the biological N cycle in tropical forest soils

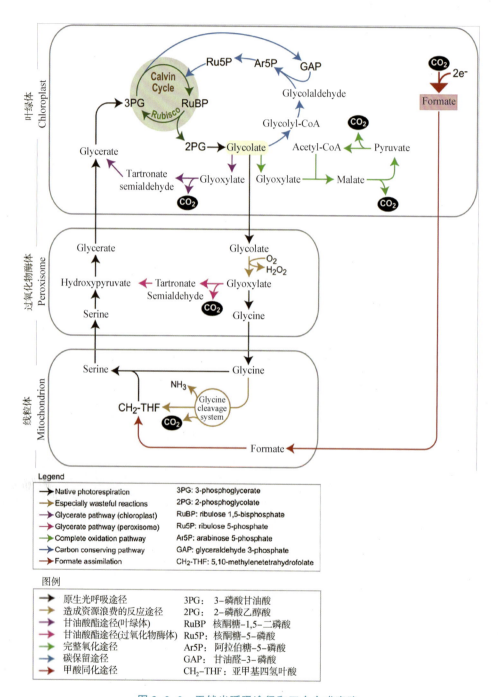

图 3-2-2 天然光呼吸途径和五个合成旁路
Fig. 3-2-2 Natural photorespiration and its synthetic bypasses

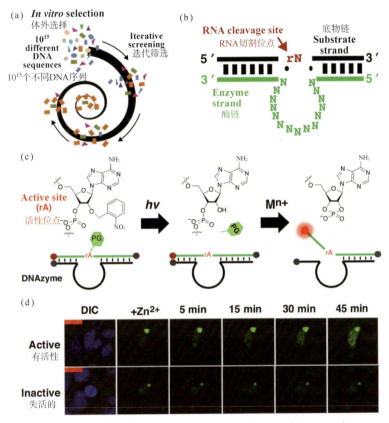

图 4-3-10 通过体外选择的迭代组合选择策略选择 DNA 酶

Fig. 4-3-10 DNAses are selected via an iterative combinatorial selection strategy called *in vitro* selection

(a) DNA 酶通过体外选择的迭代组合选择策略选择；(b) 具有根据 Watson-Crick 碱基配对杂交的结合臂 DNA 酶；
(c) DNA 酶被转变成一个催化信标；(d) 光敏在代信标的有效示例

(a) DNAses are selected via an iterative combinatorial selection strategy called *in vitro* selection;
(b) The resulting DNase has binding arms that hybridize according to Watson-Crick base pairing;
(c) DNAse turned into a catalytic beacon; (d) An example of the efficacy of the photocaged catalytic beacon

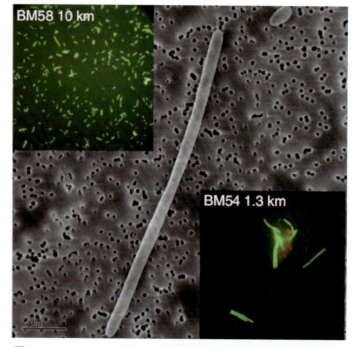

图 4-3-13　1 099~1 219 m 的显性细菌和离源头很远的橘红色斑

Fig. 4-3-13　Dominant bacteria at 1,099 to 1,219 m and acridine orange stain (inset) with distance from source

图 4-3-19　黄粉虫啃食聚苯乙烯

Fig. 4-3-19　Yellow mealworms are eating polystyrene